What Is Random?

What Is Random?

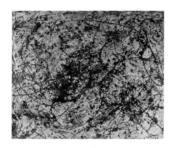

chance and order
in
mathematics
and life

Edward Beltrami

COPERNICUS
AN IMPRINT OF SPRINGER-VERLAG

©1999 Springer-Verlag New York, Inc.

On the cover: "Reflection of the Big Dipper" by Jackson Pollack.
Credit: Stedelijk Museum, Amsterdam, Holland/Superstock.

All rights reserved. No part of this publication may be reproduced, stored
in a retrieval system, or transmitted, in any form or by any means, electronic,
mechanical, photocopying, recording, or otherwise, without the prior written
permission of the publisher.

Published in the United States by Copernicus, an imprint of
Springer-Verlag New York, Inc.

Copernicus
Springer-Verlag New York, Inc.
175 Fifth Avenue
New York, NY 10010

Library of Congress Cataloging-in-Publication Data
Beltrami, Edward J.
   What is random? : chance and order in mathematics and life /
Edward Beltrami.
     p.   cm.
   Includes bibliographical references and index.
   ISBN 0-387-98737-1 (alk. paper)
   1. Probabilities.   2. Chance.   I. Title.
QA273.B395   1999
519.2—DC21   99-18389
                                                                       CIP

Manufactured in the United States of America.
Printed on acid-free paper.

9 8 7 6 5 4 3 2 1

ISBN 0-387-98737-1   SPIN 10708901

Contents

[1] The Taming of Chance 1

From Unpredictable to Lawful 2
Probability 9
Order in the Large 14
The Normal Law 18
Is It Random? 25
More About the Law of Large Numbers 30
Where We Stand Now 33

[2] Uncertainty and Information 35

Messages and Information 36
Entropy 40
Messages, Codes, and Entropy 44
Approximate Entropy 53

Contents

Again, Is It Random? 58
The Perception of Randomness 62

[3] Janus-Faced Randomness 65

Is Determinism an Illusion? 66
Generating Randomness 77
Janus and the Demons 82

[4] Algorithms, Information, and Chance 91

Algorithmic Randomness 92
Algorithmic Complexity and Undecidability 102
Algorithmic Probability 109

[5] The Edge of Randomness 117

Between Order and Disorder 118
Self-Similarity and Complexity 129
What Good is Randomness? 143

Sources and Further Readings 145

Technical Notes 153

Appendix A: Geometric Sums 179

Appendix B: Binary Numbers 183

Appendix C: Logarithims 189

References 191

Index 197

A Word About Notation

It is convenient to use a shorthand notation for certain mathematical expressions that appear often throughout the book. For any two numbers designated as a and b the product "a times b" is written as ab and sometimes as a·b, while "a divided by b" is a/b. The product of "a multiplied by itself b times" is a^b, so that, for example, 2^{10} means 1024. The expression 2^{-n} is synonymous with $1/2^n$.

In a few places I use the standard notation $\sqrt{a}$ to mean "the square root of a," as in $\sqrt{25} = 5$.

"All numbers greater than a and less than b" is expressed as (a, b). If "greater than" is replaced by "greater than or equal to" then the notation is [a, b).

A Word About Notation

A sequence of numbers such as 53371 . . . is generally indicated by $a_1\ a_2\ a_3 \ldots$, in which the subscripts 1, 2, 3, . . . indicate the "first, second, third, and so on" terms of the sequence, which can be finite in length or even infinite (such as the unending array of all even integers).

Preface

We all have memories of peering at a TV screen as myriad little balls churn about in an urn until a single candidate, a number inscribed on it, is ejected from the container. The hostess hesitantly picks it up and, pausing for effect, reads the lucky number. The winner thanks Lady Luck, modern descendant of the Roman goddess Fortuna, blind arbiter of good fortune. We, as spectators, recognize it as simply a random event and know that in other situations Fortuna's caprice could be equally malicious, as Shirley Jackson's dark tale "The Lottery" chillingly reminds us.

The *Oxford Dictionary* has it that a "random" outcome is one without perceivable cause or design, inherently unpredictable. But, you might protest, isn't the world around us

Preface

governed by rules, by the laws of physics? If that is so, it should be possible to determine the positions and velocities of each ball in the urn at any future time, and the uncertainty of which one is chosen would simply be a failing of our mind to keep track of how the balls are jostled about. The sheer enormity of possible configurations assumed by the balls overwhelms our computational abilities. But this would be only a temporary limitation: A sufficiently powerful computer could conceivably do that brute task for us, and randomness would thus be simply an illusion that can be dispelled. After thinking about this for a while you may begin to harbor a doubt. Although nature may have its rules, the future remains inherently unknowable because the positions and velocities of each ball can never really be ascertained with complete accuracy.

The first view of randomness is of clutter bred by complicated entanglements. Even though we know there are rules, the outcome is uncertain. Lotteries and card games are generally perceived to belong to this category. More troublesome is that nature's design itself is known imperfectly, and worse, the rules may be hidden from us, and therefore we cannot specify a cause or discern any pattern of order. When, for instance, an outcome takes place as the confluence of totally unrelated events, it may appear to be so surprising and bizarre that we say that it is due to blind chance. Jacques Monod, in his book *Chance and Necessity,* illustrates this by the case of a man hurrying down a street in response to a sudden phone call at the same time that a roof worker accidentally drops a hammer that hits the unfortunate pedestrian's head. Here we have chance

Preface

due to contingency, and it doesn't matter whether you regard this as an act of divine intervention operating according to a predestined plan or as an unintentional accident. In either case the cause, if there is one, remains indecipherable.

Randomness is the very stuff of life, looming large in our everyday experience. Why else do people talk so much about the weather, traffic, and the financial markets? Although uncertainty may contribute to a sense of anxiety about the future, it is also our only shield against boring repetitiveness. As I will argue in the final chapter, chance provides the fortuitous accidents and capricious wit that give life its pungency. It is important, therefore, to make sense of randomness beyond its anecdotal meanings. To do this we employ a modest amount of mathematics in this book to remove much of the vagueness that encumbers the concept of random, permitting us to quantify what would otherwise remain elusive. The mathematics also provides a framework for unifying our understanding of how chance is interpreted from the diverse perspectives of psychologists, physicists, statisticians, computer scientists, and communication theorists.

In the first chapter I tell the story of how, beginning a few centuries ago, the idea of uncertainty was formalized into a theory of chance events, known today as probability theory. Mathematicians adopt the convention that selections are made from a set of possible outcomes in which each event is equally likely, though unpredictable. Chance is then asked to obey certain rules that epitomize the behavior of a perfect coin or an ideal die, and from these rules one can calculate the odds.

Preface

"The Taming of Chance," as this chapter is called, supplies the minimal amount of probability theory needed to understand the "law of large numbers" and the "normal law," which describe in different ways the uncanny regularity of large ensembles of chance events. With these concepts in hand, we arrive at our first tentative answer to the question "what is random?"

To crystalize our thinking, most of the book utilizes the simplest model of a succession of random events, namely a binary sequence (a string of zeros and ones). Although this may seem almost like a caricature of randomness, it has been the setting for some of the most illuminating examples of the workings of chance from the very beginnings of the subject three centuries ago to the present.

If some random mechanism generates a trail of ten zeros and ones, let us say, then there are 2 to the power 10, namely 1024, different binary strings that are possible. One of the first conundrums to emerge is that under the assumption that each string is equally likely, there is a (very) small possibility that the string generated consists of ten zeros in succession, something that manifestly is not random by any intuitive notion of what random means. So it is necessary to distinguish between a device that operates in a haphazard manner to spew forth digits without rhyme or reason, a random process, and any particular realization of the output that such a process provides. One may argue that the vagaries of chance pertain to the ensemble of possibilities and not to an individual outcome. Nevertheless, a given binary string, such as 0101010101010101, may appear so orderly that we feel compelled to deny its random-

Preface

ness, or it may appear so incoherent, as in the case of 0011000010011011, that we yearn to call it random regardless of its provenance. The last situation conforms to what I mentioned earlier, namely that random is random even though the cause, if any, is unknown. In keeping with this idea the emphasis will shift away from the generating process in succeeding chapters and focus instead on the individual outcomes. This change in outlook parallels the actual shift that has taken place in the last several decades among a broad spectrum of thinkers on the subject, in contrast to the more traditional views that have been central over the last few centuries (and that are discussed in the first chapter).

Entertaining puzzles and surprising inferences sometimes spring from chance in unexpected ways, and they spotlight the often counterintuitive consequences of randomness in everyday life. Although there are a few examples of this kind in the first chapter, this is not the focus of the present work (the splendid book of William Feller mentioned as reference [27] provides a host of such examples). My goal, instead, is to unearth the footprints of randomness rather than to show how randomness, per se, can confound us.

In "Uncertainty and Information," the second chapter, I introduce the notion of information formulated by Claude Shannon a half century ago, since this gives us an additional tool for discussing randomness. Here one talks of information bits and entropy, redundancy and coding, and this leads me to pose a second test to resolve the question "what is random?" The idea is that if some shorter binary string can serve as a code

Preface

to generate a longer binary message, the longer message cannot be random, since it has been compressed by removing some of the redundancies within the string. One application concerns the perception of randomness by people in general, a topic much studied by psychologists.

The third chapter, "Janus-Faced Randomness," introduces a curious example of a binary sequence that is random in one direction but deterministic when viewed in reverse. Knowledge of the past and uncertainty about the future seem to be two faces of the same Janus-faced coin, but more careful scrutiny establishes that what appears as predictable is actually randomness in disguise. I establish that the two faces represent a tradeoff between ignorance now and disorder later. Though there are randomly generated strings that do not appear random, it now emerges that strings contrived by deterministic rules may behave randomly. To some extent randomness is in the eye of the beholder. Just because you do not perceive a pattern doesn't mean there isn't one. The so-called random-number generators that crop up in many software packages are of this ilk, and they give support to the interpretation of chance as a complex process—like balls in an urn, which only appear random because of the clutter.

The chapter concludes with a brief discussion of the early days of the study of thermodynamics in the nineteenth century, when troublesome questions were raised about the inexorable tendency of physical systems to move from order to disorder over time, in apparent contradiction to the laws of physics, which are completely reversible. I show that this

Preface

dilemma is closely related to the question of binary strings that appear random in spite of being spawned by precise rules.

In the fourth chapter, "Algorithms, Information, and Chance," I follow the mathematicians Andrei Kolmogorov and Gregory Chaitin, who say that a string is random if its shortest description is obtained by writing it out in its entirety. There is, in effect, no pattern within the string that would allow it to be compressed. More formally, one defines the complexity of a given string of digits to be the length, in binary digits, of the shortest string (that is, the shortest computer program written in binary form) that generates the successive digits. Strings of maximum complexity are called random when they require programs of about the same length as the string itself. Although these ideas echo those of Chapter 2, they are now formulated in terms of algorithms implemented on computers, and the question "what is random?" takes on a distinctly different cast.

Gödel's celebrated incompleteness theorem in the version rendered by Alan Turing states that it may not be possible to determine whether a "universal" computer, namely one that can be programmed to carry out any computation whatever, will ever halt when it is fed a given input. Chaitin has reinterpreted this far-reaching result by showing that any attempt to decide the randomness of a sufficiently long binary string is inherently doomed to failure; his argument is reproduced in this chapter.

The last chapter is more speculative. In "The Edge of Randomness" I review recent work by a number of thinkers

Preface

that suggests that naturally occurring processes seem to be balanced between tight organization, where redundancy is paramount, and volatility, in which little order is possible. One obtains a view of nature and the arts and the world of everyday affairs as evolving to the edge of these extremes, allowing for a fruitful interplay of chance and necessity, poised between surprise and inevitability. Fortuitous mutations and irregular natural disturbances, for example, appear to intrude on the more orderly processes of species replication, providing evolution with an opportunity for innovation and diversity.

To illustrate this concept in a particular setting, I include a brief and self-contained account of naturally occurring strings of numbers known as fractals that display similar patterns at different scales. Although the algorithmic complexity of these strings may be substantial, they are not entirely random, since many of them describe processes in which the frequencies of events, from small to large, exhibit a regular pattern. This is reminiscent of the order exhibited by chance events in the large as discussed in the first chapter. However, the prediction of any individual event in a fractal string remains inscrutable because it is contingent on the past history of the process.

The book is intended to provoke, entertain, and inform by challenging the reader's ideas about randomness, providing first one and then another interpretation of what this elusive concept means. As the book progresses I tease out the various threads and show how mathematics, communication engineering, computer science, philosophy, physics, and psychol-

Preface

ogy all contribute to the discourse by illuminating different facets of the same idea.

The material in the book should be readily accessible to anyone with a smattering of college mathematics, *no calculus needed*. I provide simple numerical examples throughout, coded in MATLAB, to illustrate the various iterative processes and binary sequences that crop up. Three appendices provide some of the background information regarding binary representations and logarithms that are needed here and there, but I keep this as elementary as possible. Although an effort is made to justify most statements of a mathematical nature, a few are presented without corroboration, since they entail close-knit arguments that would detract from the main ideas. You can safely bypass the details without any loss, and in any case, the fine points are available in the technical notes assembled at the end. Reference to a technical note is indicated by a superscript.

Acknowledgments

This book is an amalgam of many sources and ideas. It began to take shape some years ago, when I first became aware of a paper by the statistician M. Bartlett that led to the conundrum of the Janus iterates discussed in Chapter 3 and 4. More recently, Professor Laurie Snell, of Dartmouth College, kindly provided me with some of his commentary from the Chance website, and this stimulated me to put together a set of notes on randomness that ultimately became the present book. Incidentally, I strongly recommend this website (www.dartmouth .edu/~chance) as a valuable resource for teachers and the general reader, since it covers a wide range of topics connected to the appearance of chance in everyday life.

Acknowledgments

I wish to thank Jonathan Cobb, formerly Senior Editor of Copernicus Books at Springer-Verlag, New York, and Dr. David Kramer for a detailed and very helpful commentary on the entire manuscript. Their editorial expertise is greatly appreciated. I am also grateful to Alexis Beltrami, the prototypical educated layman, for his critical comments on the last chapter, and my thanks to Professor Hondshik Ahn, of the State University at Stony Brook, for a critical review of the the first chapter.

Finally, I am indebted to my wife, Barbara, for providing sound advice that led to improvements in garbled early versions of the book, and for lending a patient ear to my endless chatter about randomness during our frequent walks together.

[1]

The Taming of Chance

An oracle was questioned about the mysterious bond between two objects so dissimilar as the carpet and the city . . . for some time augurs had been sure that the carpet's harmonious design was of divine origin . . . but you could, similarly, come to the opposite conclusion: that the true map of the universe is the city, just as it is, a stain that spreads out shapelessly, with crooked streets, houses that crumble one upon the other amid clouds of dust.

Italo Calvino, *Invisible Cities*

What Is Random?

However unlikely it might seem, no one had tried out before then a general theory of chance. Babylonians are not very speculative. They revere the judgments of fate, they deliver to them their lives, their hopes, their panic, but it does not occur to them to investigate fate's labyrinthine laws nor the gyratory spheres which reveal it. Nevertheless . . . the following conjecture was born: if the lottery is an intensification of chance, a periodical infusion of chaos in the cosmos, would it not be right for chance to intervene in all stages of the drawing and not in one alone? Is it not ridiculous for chance to dictate someone's death and not have the circumstances of that death—secrecy, publicity, the fixed time of an hour or a century—not subject to chance?

Jorge Luis Borges, "The Lottery in Babylon"

From Unpredictable to Lawful

In the dim recesses of ancient history the idea of chance was intertwined with that of fate. What was destined to be, would be. Chance was personified, in the Roman Empire at least, by the Goddess Fortuna, who reigned as the sovereign of cynicism and fickleness. As Howard Patch puts it in his study of this Roman deity, "to men who felt that life shows no signs of fair-

The Taming of Chance

ness, and that what lies beyond is at best dubious, that the most you can do is take what comes your way, Fortuna represented a useful, if at times flippant, summary of the way things go."

To subvert the willfulness of Fortuna one could perhaps divine her intentions by attempting to simulate her own mode of behavior. This could be accomplished by engaging in a game of chance, in hope that its results would reveal what choice Fortuna herself would make. There is evidence that pre-Christian people along the Mediterranean coast tossed animal heel bones and that this eventually evolved into play with dice. When a chance outcome out of many possible outcomes had been revealed by the casting of lots, one could then try to interpret its omens and portents and decide what action to take next. Julius Caesar, for instance, resolved his agonizing decision to cross the Rubicon and advance upon Rome by allegedly hurling dice and exclaiming *"iacta alea est,"* the die is cast. This is echoed in our own time when people draw lots to decide who will be first or, for that matter, who is to be last. The very essence of fair play is to flip a coin so that chance can decide the next move.

The interpretation of omens was an attempt to decipher the babble of seemingly incoherent signs by encoding them into a compact and coherent prophecy or, perhaps, as a set of instructions to guide the supplicant. Seen this way, divination is a precursor to the use of coding in information theory and message compression in algorithmic complexity theory, topics that will figure prominently in later portions of this book.

What Is Random?

Fortuna was not all foreboding. There is evidence that the elements of chance were employed not just for augury and divination but, on a more playful side, for diversion. Games involving chance and gambling were firmly established by the Renaissance, and a number of individuals had begun to notice that when regularly shaped and balanced dice were tossed repeatedly, certain outcomes, such as five dots on one of the six faces, seemed to occur on the average about a sixth of the time. No one was more persuasive about this than the sixteenth-century figure Girolamo Cardano, a celebrated Italian physician and mathematician, who wrote a small tract on gambling, *Liber de Ludo Aleae,* in which he demonstrated his awareness of how to calculate the winning odds in various games.

This newly found grasp on the workings of chance in simple games would soon evolve into a broader understanding of the patterns revealed by chance when many observations are made. In retrospect, it seems inevitable that the study of randomness would turn into a quantitative science, parallel to the manner in which the physical sciences were evolving during the late Renaissance. Over the next two centuries a number of individuals, such as the mathematicians Blaise Pascal and Pierre de Fermat, had a hand in reining in the arbitrary spirit of Fortune, at least in games of dice and cards, but it wasn't until the early eighteenth century, after the calculus had been invented and mathematics in general had reached a level of maturity unthinkable a few centuries earlier, that probable and improbable events could be adequately expressed in a mathematical form. It was at this time that the tools were forged that led to the modern theory of probability. At the same time the study of statistics as we now know it began to emerge from the data-gathering efforts directed at providing mortality tables and insurance annuities, both of which hinge on chance events.

The Swiss mathematician Jakob Bernoulli, in his *Ars Conjectandi* of 1713, and, shortly thereafter, Abraham de Moivre, in *The Doctrine of Chances,* stripped the element of randomness to its bare essentials by considering only two possible outcomes, such as black and white balls selected blindly from an urn, or tosses of an unbiased coin. Imagine that someone flips a balanced coin a total of n times, for some integer n. The proportion of heads in these n tosses, namely, the actual number of heads produced divided by n, is a quantity usually referred to as the *sample average,* since it depends on the particular sample of a sequence of coin flips that one gets. Different outcomes of n tosses will generally result in different sample averages.

What Bernoulli showed is that as the sample size n gets larger, it becomes increasingly likely that the proportion of heads in n flips of a balanced coin (the sample average) will not deviate from one-half by more than some fixed margin of error. This assertion will be rephrased later in more precise terms as the "Law of Large Numbers." A few years later, de Moivre put more flesh on Bernoulli's statement by establishing that if the average number of heads is computed many times, most of the sample averages have values that cluster about ½, while the remainder spread themselves out more sparsely the further one gets from ½. Moreover, de Moivre

What Is Random?

showed that as the sample size gets larger, the proportion of sample averages that are draped about ½ at varying distances begins to look like a smooth bell-shaped curve, known either as the *normal* or the *Gaussian* curve (after the German mathematician Carl Friedrich Gauss, whose career straddled the eighteenth and nineteenth centuries). This phenomenon is illustrated in Figure 1.1, in which 10,000 sample averages are grouped into little segments along the horizontal axis. The height of the rectangular bins above the segments represents the number of sample averages that lie within the indicated interval. The rectangles decrease in size the further one gets from ½, which shows that fewer and fewer sample averages are to be found at longer distances from the peak at ½. The overall profile of the rectangles, namely the distribution of sample averages, is bell-shaped, and as n increases, this profile begins to approximate ever more closely the smooth (Gaussian) curve that you see superimposed on the rectangles in the figure. Though formal proofs of de Moivre's theorem and of Bernoulli's law are beyond the scope of the present work, the results themselves will be put to good use later in this chapter.

Taken together, the assertions of Bernoulli and de Moivre describe a kind of latent order emerging from a mass of disordered data, a regularity that manifests itself amid the chaos of a large sample of numbers. By the early years of the nineteenth century, the use of the new methods to handle large masses of data were enthusiastically employed to harness uncertainty in every sphere of life, beyond the fairly benign examples of games, and never more so than by zealous government bu-

The Taming of Chance

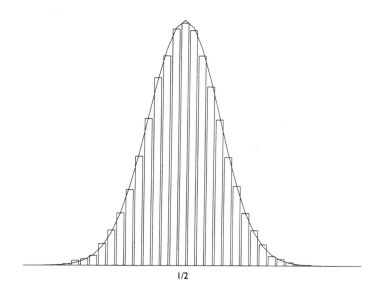

Figure 1.1 The distribution of 10,000 sample averages at varying distances from ½. The smooth bell-shaped curve is a ubiquitous presence in statistics and is known as the normal, or sometimes the Gaussian, curve.

reaucracies bent on grappling with the torrents of data being collected about the populations within their borders. Insanity, crime, and other forms of aberrant behavior had been abundantly cataloged in the eighteenth and early nineteenth centuries, and a way for extracting the social implications of these numbers had at last become available. One could now define "normal" behavior as being within a certain fraction of the

What Is Random?

average of a large sample of people, with deviants lying outside this range, in the same way as one spoke of the average of many coin tosses.

What was once puzzling and unpredictable now appeared to fall into patterns dictated by the normal curve. This curve was empirically regarded by sciences like astronomy as a "law of errors," but some scientists, the French astronomer Adolphe Quetelet in particular, turned it into a bastion of social theory. By 1835 Quetelet went so far as to frame the concept of "the average man" that is still with us today. The period of sorting and interpreting data, the very stuff of statistics, had begun in earnest, and it is no exaggeration to say that the taming of chance had come of age.

Throughout the nineteenth and into the beginnings of the twentieth century, the applications of mathematical ideas to the study of uncertainty became more widespread, and in particular, they played an important role in physics with the study of statistical mechanics (a topic that we touch on in Chapter 3), where large ensembles of molecules collide with each other and scatter at random.

The mathematical methods for the study of random phenomena became more sophisticated throughout the first part of the twentieth century, but it remained for the Russian mathematician Andrei Kolmogorov to formalize the then current thinking about probability in a short but influential monograph that was published in 1933. He established a set of hypotheses about random events that when properly used could explain how chance behaves when one is confronted with a large number of similar observations of some phenomenon. Probability has since provided the theoretical underpinnings of statistics, in which inferences are drawn from numerical data by quantifying the uncertainties inherent in using a finite sample to draw conclusions about a very large or even unlimited set of possibilities. You will see a particular instance of statistical inference a little later in this chapter, when we attempt to decide whether a particular string of digits is random.

The fascination of randomness is that it is pervasive, providing the surprising coincidences, bizarre luck, and unexpected twists that color our perception of everyday events. Although chance may contribute to our sense of unease about the future, it is also, as I argue in the final chapter, a bulwark against stupefying sameness, giving life its sense of ineffable mystery. That is why we urgently ask, "What is random?"

The modest amount of mathematics employed in this book allows us to move beyond the merely anecdotal meanings of randomness, quantifying what would otherwise remain elusive and bringing us closer to knowing what random is.

Probability

The subject of probability begins by assuming that some mechanism of uncertainty is at work giving rise to what is called randomness, but it is not necessary to distinguish between chance that occurs because of some hidden order that may exist and chance that is the result of blind lawlessness. This mechanism, figuratively speaking, churns out a succession of events, each individually unpredictable, or it conspires to produce an unforeseeable outcome each time a large ensemble of possibilities is sampled.

A requirement of the theory of probability is that we be able to describe the outcomes by numbers. The totality of deaths from falling hammers, the winning number in a lottery, the price of a stock, and even yes and no situations, as in it does or does not rain today (which is quantifiable by a zero for "no" or a one for "yes"), are all examples of this.

The collection of events whose outcomes are dictated by chance is called a *sample space* and may include something as simple as two elements, head or tail in coin tossing, or it may be something more involved, as the eight triplets of heads and tails in three tosses, or the price of a hundred different stocks at the end of each month. The word "outcome" is interchangeable with observation, occurrence, experiment, or trial, and in all cases it is assumed that the experiment, observation, or whatever can be repeated under essentially identical conditions as many times as desired even though the outcome each time is unpredictable. With games, for example, one assumes that cards are thoroughly shuffled, or that balls scrambled in an urn are selected by the equivalent of a blindfolded hostess. In other situations, where the selection mechanism may not be under our control, it is assumed that nature arbitrarily picks one of the possible outcomes. It matters not whether we think of the succession of outcomes as a single experiment repeated

The Taming of Chance

many times or a large sample of many experiments carried out simultaneously. A single die tossed three times is the same as three dice tossed once each.

Each possible outcome in a finite sample space is called an *elementary event* and is assigned a number in the range from zero to one. This number, the event's *probability,* designates the likelihood that the event takes place. Zero means that it cannot happen, and the number one is reserved for something that is certain to occur. The interesting cases lie in between.

Imagine that a large, possibly unlimited, number of observations are made and that some particular event takes place unexpectedly from time to time. The probability of this event is a number between zero and one that expresses the ratio between the actual number of occurrences of the event and the total number of observations. In tossing a coin, for example, head and tail are each assigned a probability of ½ whenever the coin seems to be balanced. This is because one expects that the event of a head or tail is equally likely in each flip, and so the average number of heads (or tails) in a large number of tosses should be close to ½.

More general events in a sample space are obtained by considering the union of several elementary events. A general event E (sometimes other letters are used, such as A or B) has a probability assigned to it just as was done with the elementary events. For example, the number of dots in a single toss of a balanced die leads to six elementary events, namely the number of dots, from one to six, on the upturned face. Event E might then be "the upturned face shows a number greater than four," which is

What Is Random?

the union of the elementary events "5 dots" and "6 dots," and in this case, the probability of "5 or 6 dots" is $2/6$, or $1/3$.

Two events that represent disjoint subsets of a sample space, subsets having no point in common, are said to be *mutually exclusive*. If A and B represent mutually exclusive events, the event "either A or B takes place" has *a probability equal to the sum of the individual probabilities of A and B*. The individual probabilities of the union of any number of mutually exclusive events (namely, a bunch of events that represent sets of possibilities having nothing in common) must add up to unity since one of them is certain to occur.

If the occurrence or nonoccurrence of a particular event is hit or miss, totally unaffected by whether it happened before, we say that the outcomes are *statistically independent* or, simply, *independent*.

For a biased coin in which a head has only probability $1/3$ of happening, the long-term frequency of heads in a sequence of many tosses of this coin carried out in nearly identical circumstances should appear to settle down to the value $1/3$. Although this probability reflects the uncertainty associated with obtaining a head or tail for a biased coin, I attach particular significance in this book to unbiased coins or, more generally, to the situation in which all the elementary events are equally likely. When there is no greater propensity for one occurrence to take place over another, the outcomes are said to be *uniformly distributed,* and in this case, the probabilities of the N separate elementary events that constitute some sample space are each

The Taming of Chance

equal to the same value $1/N$. For instance, the six faces of a balanced die are all equally likely to occur in a toss, and so the probability of each face is $\frac{1}{6}$.

The quintessential *random process* will be thought of, for now at least, as a succession of *independent and uniformly distributed outcomes*.

To avoid burdening the reader at this point with what might be regarded as boring details, I append additional information and examples about probability in an optional section entitled "More About Probability" in the technical notes for this chapter.[1.1]

In thinking about random processes in this book it simplifies matters if we regard successive outcomes as having only two values, zero or one. This could represent any binary process such as coin tossing or, in more general circumstances, occurrence or nonoccurrence of some event (success or failure, yes or no, on or off). Ordinary whole numbers can always be expressed in *binary notation,* namely a string of zeros and ones. It suffices to write each number as a sum of powers of 2. For example,

$$9 = 1 \cdot 2^3 + 0 \cdot 2^2 + 0 \cdot 2^1 + 1 \cdot 2^0 = 8 + 0 + 0 + 1;$$

the number 9 is then identified with the binary digits multiplying the individual powers of 2, namely 1001. Similarly,

$$30 = 1 \cdot 2^4 + 1 \cdot 2^3 + 1 \cdot 2^2 + 1 \cdot 2 + 0 \cdot 2^0$$
$$= 16 + 8 + 4 + 2 + 0,$$

What Is Random?

and so 30 is identified with 11110 (additional details about binary representations are provided in Appendix B).

The examples in this chapter, and indeed throughout most of this book, are framed in terms of binary digits. Any outcome, such as the price of a volatile stock or the number of individuals infected with a contagious disease, may therefore be coded as a string of zeros and ones called *binary strings* or, as they are sometimes referred to, *binary sequences*.

Computers, incidentally, express numbers in binary form, since the circuits imprinted on chips respond to low/high voltages, and the familiar Morse code, long the staple of telegraphic communication, operates with a binary system of dots and dashes.

Order in the Large

It has already been mentioned that one expects the proportion of heads in n flips of a balanced coin to be close to $\frac{1}{2}$ leading one to infer that the probability of a head (or tail) is precisely $\frac{1}{2}$. What Jakob Bernoulli did was to turn this assertion into a formal mathematical theorem by hypothesizing a sequence of independent observations, or, as they are often referred to, independent trials, of a random process with two possible outcomes that are labeled zero or one, each mutually exclusive of the other. He assumed an exact probability, designated by $p,$ for the event "one will occur" and a corresponding probability $1-p$ for the event "zero will occur" (namely, the digit one does not appear). Since one of the two events must take place at each trial, their probabilities p and $1-p$ sum to unity.

These assumptions idealize what would actually be observed in viewing a binary process, in which there are only two outcomes. The sequence of outcomes, potentially unlimited in number, will be referred to as *Bernoulli* p-*trials* or, equivalently, as a *Bernoulli p-process.*

In place of one and zero, the binary outcomes could also be thought of as success and failure, true and false, yes and no, or some similar dichotomy. In medical statistics, for example, one might ask whether or not a particular drug treatment is effective when tested on a group of volunteers selected without bias from the population at large.

For a sequence of *Bernoulli* p-*trials* you intuitively expect that the proportion of ones in n independent trials should approach p as n increases. In order to turn this intuition into a more precise statement, let S_n denote the number of ones in n trials. Then S_n divided by n, namely S_n/n, represents the fraction of ones in n trials; it is customary to call S_n/n the *sample average*. What Jakob Bernoulli established is that as n increases it becomes increasingly certain that the absolute difference between S_n/n and p is less than any preassigned measure of discrepancy. Stated more formally, this says that the probability of the event "S_n/n is within some fixed distance from ½" will tend to one as n is allowed to increase without bound. Bernoulli's theorem is known as the *Law of Large Numbers.*

In the familiar toss of a balanced coin, where p equals ½, we expect the sample average to be close to ½ in a long sequence of

What Is Random?

flips, and Bernoulli's Law of Large Numbers gives substance to our intuition that an unknown probability p can be estimated a posteriori by computing S_n/n for large n. After a large number of independent trials have taken place it is therefore nearly correct to say that *the "probability p" is equal to the "proportion (or percentage) of successes,"* and the two terms in quotation marks are often used synonymously. More generally, if an event E occurs r times in n independent trials, the probability of E is very nearly equal to the relative frequency r/n for sufficiently large n.

The Bernoulli ½-trials are the quintessential example of a random process as we defined it earlier: The succession of zeros and ones is independent and uniformly distributed with each digit having an equal chance of occurring.

Although the Law of Large Numbers for a Bernoulli ½-process states that the probability of the event "the sample average S_n/n deviates from ½ by less than some small preassigned value" can be made as close to 1 as we wish by choosing n big enough, it leaves open the possibility that S_n/n will fail to be close to ½ in subsequent trials. The values of S_n/n can fluctuate considerably, and our only consolation is that these aberrant excursions from the vicinity of ½ will occur less and less frequently as n increases. In Figure 1.2 there is a plot of S_n/n versus n generated by a computer-simulated random coin toss with $p = ½$, and in this instance we see that the sample mean does meander towards ½ in its own idiosyncratic manner.

This is a good a place to comment on the oft-quoted "*law of averages,*" which is epitomized by the belief that after a long streak of bad luck, as seen by a repeated block of zeros (tails,

The Taming of Chance

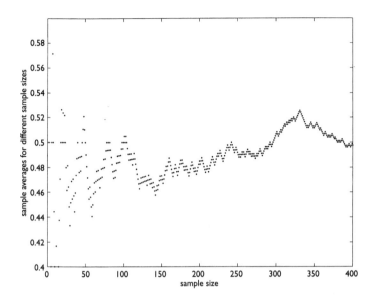

Figure 1.2 Fluctuations in the value of the sample average S_n/n for n up to 400. Note that the fluctuations appear to settle down to the value ½ as n gets larger, much as the Law of Large Numbers leads us to expect.

you lose, for example), fortune will eventually turn things around by favoring the odds of a one (heads, or a win). The perception is that a long string of zeros is so unlikely that after such a bizarre succession of losses the chance of a head is virtually inevitable, and a gambler has a palpable sense of being on the cusp of a win. However, the independence of tosses exposes this as a delusion: The probability of another zero remains the

17

same ½ as for every other toss. What is true is that in the ensemble of all possible binary strings of a given length the likelihood that any particular string has a freakishly long block of zeros is quite small. This dichotomy between what is true of a particular realization of a random process and the ensemble of all possible outcomes is a persistent thorn in the flank of probability theory.

The "law of averages" surfaces again in the form of an intuition that in a game of heads and tails, heads will be in the lead roughly half the time, with a win (heads) for the most part compensating a loss (tails) throughout. But this is dead wrong. Contrary to expectation, it is likely that either heads or tails will be in the lead throughout the game. In Figure 1.3 accumulated winnings of 5000 tosses are plotted, and it illustrates dramatically that *a lead or loss is maintained for most of the game,* even though it tends to fluctuate in value. This does not contradict the Law of Large Numbers, which implies that in a large number of different coin tossing games "heads" will be in the lead roughly one-half the time; likewise for tails.

The Normal Law

As we mentioned earlier, de Moivre gave a new twist to the Law of Large Numbers for Bernoulli p-trials. The Law of Large Numbers says that for a large enough sample size, the sample average is likely to be near p, but de Moivre's theorem allows one to estimate the probability that a sample average is

The Taming of Chance

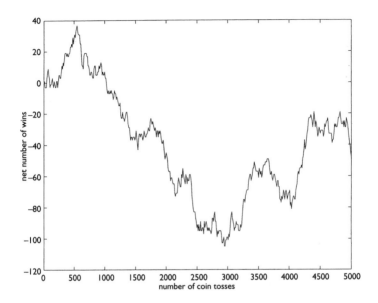

Figure 1.3 Fluctuations of total wins and losses in a game of heads and tails with a fair coin over a span of 5000 tosses, plotted every tenth value. A loss is clearly maintained in over more than 4000 flips of the coin.

within a specified distance from p. One first must choose a measure of how far S_n/n is from p. This measure is conventionally defined to be $\sigma/\sqrt{n}$, where the symbol σ stands for $\sqrt{p(1-p)}$. With p equal to .25, for example, σ is equal to $\sqrt{3/16}$, or $\sqrt{3}/4$.

De Moivre's theorem says that the proportion (percentage) of sample averages S_n/n that fall in the so-called *confidence*

What Is Random?

interval of values between $p - c\,\sigma/\sqrt{n}$ and $p + c\,\sigma/\sqrt{n}$ for a given value of c is approximately equal to the fraction of total area that lies under the normal curve between $-c$ and c. The *normal curve,* often called the *Gaussian curve,* or even the *law of errors,* is the ubiquitous bell-shaped curve that is familiar to all students of statistics and that is displayed in Figure 1.1. The essence of de Moivre's theorem is that the error made in using the normal curve to approximate the probability that S_n/n is within the given confidence interval decreases as n gets larger. It is this fact that will shortly allow us to devise a test, the first of several, for deciding whether a given binary string is random or not.

The area under the normal curve has been computed for all c, and these numbers are available as a table in virtually every statistics text and statistics software package. One value of c that is particularly distinguished by common usage is 1.96, which corresponds, according to the tables, to a probability of .95. A probability of .95 means that S_n/n is expected to lie within the confidence interval in 19 out of 20 cases or, put another way, the odds in favor of S_n/n falling within the interval are 19 to 1.

In preparation for using the above results to test randomness, I will attempt to put de Moivre's theorem in focus for p = ½ (think of a balanced coin that is not biased in favor of falling on one side or the other). Before we carry out this computation, I hope you will agree that it is less cumbersome to express the confidence interval by using the mathematical shorthand

$$\left(p - \frac{c\sigma}{\sqrt{n}},\ p + \frac{c\sigma}{\sqrt{n}}\right)$$

Since the quantity σ is the square root of ¼ in the present case, namely ½, and since a probability of .95 corresponds to $c = 1.96$, as you just saw, the 95% *confidence interval for the sample average is*

$$\left(.5 - \frac{1.96}{2\sqrt{n}},\ .5 + \frac{1.96}{2\sqrt{n}}\right)$$

By rounding 1.96 up to 2.0, the confidence interval takes on the pleasingly simple form

$$\left(.5 - \frac{1}{\sqrt{n}},\ .5 + \frac{1}{\sqrt{n}}\right)$$

which means that roughly 95% of the sample averages are expected to take on values between .5 plus or minus $1/\sqrt{n}$. For example, in 10,000 tosses of a coin, the 95% confidence interval is (.49, .51), since the square root of 10,000 is 100. If there are actually 5300 heads in 10,000 tosses, then S_n/n is .53, and since we are 95% confident that S_n/n should be between .49 and .51, we are inclined to disbelieve that the coin is balanced (p equal to ½) in favor of thinking that it is unbalanced (p not equal to ½). Computations like this are part of the obligatory lore of most courses in statistics. Observe that the confidence interval narrows, for a given probability (area), in inverse proportion to $\sqrt{n}$. That is, the longer the value of n, the smaller is the interval about ½ in which we expect 95% of the sample averages to fall.

What Is Random?

The distribution of values of 10,000 different samples of S_n/n for $n = 15$ is plotted in Figure 1.4 (upper half) for the case $p = \frac{1}{2}$, and we see that they cluster about $\frac{1}{2}$ with a dispersion that appears roughly bell-shaped. The height of each rectangle in the figure indicates the number of all sample averages that lie in the indicated interval along the horizontal axis. For purposes of comparison, the actual normal curve is superimposed on the rectangles, and it represents the theoretical limit approached by the distribution of sample averages as n increases without bound.

The lower half of Figure 1.4 compares the difference between using 10,000 sample averages of size $n = 60$. Since the dispersion about $\frac{1}{2}$ decreases inversely with the square root of n, the bottom distribution is narrower and its peak is higher. The 95% confidence interval, in particular, has a smaller width when $n = 60$.

It appears, then, that from the disarray of individual numbers a certain lawfulness emerges when large ensembles of numbers are viewed from afar. The wanton and shapeless behavior of numbers when seen up close gels, in the large, into the orderly bell-shaped form of a "law of errors," or what is appropriately called the *normal law, or even the Gaussian law,* as established by de Moivre's theorem. Incidentally, the mathematician Gauss had little to do with the normal law, but his name is nonetheless attached to this result, possibly because of his influence in other areas of mathematics.

Extensions of the normal law to other than Bernoulli p-trials is known today as the "Central Limit Theorem." It

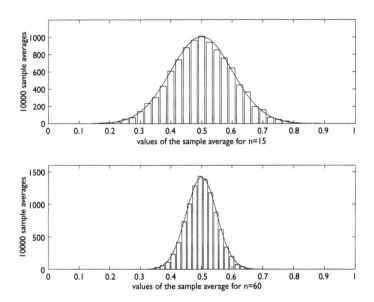

Figure 1.4 The distribution of 10,000 sample averages S_n/n for $n = 15$ in the top half. The height of each rectangle indicates the fraction of all sample averages that lie in the indicated interval of the horizontal axis. The distribution of 10,000 sample averages for $n = 60$ is on the bottom half. The spread about ½ in the bottom figure is less than in the upper, and it peaks higher. Thus, the 95% confidence interval is narrower. This is because the dispersion about ½ is inversely proportional to the square root of n. Observe that the distribution of values appears to follow a normal curve.

What Is Random?

must be remembered, however, that the normal law is a mathematical statement based on a set of assumptions about the perception of randomness that may or may not always fully square with actuality. Probability theory permits statements like the Law of Large Numbers or the Central Limit Theorem to be given a formal proof, capturing what seems to be the essence of statistical regularity as it unfolds from large arrays of messy data, but it does not, of itself, mean that nature will accommodate us by guaranteeing that the sample average will tend to p as n increases without bound or that the values of the sample average will distribute themselves according to a normal curve. No one ever conducts an infinite sequence of trials. What we have instead is an empirical observation based on limited experience, and not a law of nature. This reminds me of an oft-quoted observation of the mathematician Henri Poincaré to the effect that practitioners believe that the normal law is a theorem of mathematics, while the theoreticians are convinced that it is a law of nature. It is amusing to compare this to the ecstatic remarks of the nineteenth-century social scientist Francis Galton, who wrote, "I scarcely know of anything so apt to impress the imagination as the wonderful form of cosmic order expressed by the "Law of Frequency of Error" [*what I refer to as the normal law*]; it reigns with serenity and in complete self-effacement, amidst the wildest confusion. The larger the mob, and the greater the apparent anarchy, the more perfect is its sway. It is the supreme law of unreason."

The Taming of Chance

Is It Random?

If there were a signature example in statistics, it would be to test whether a particular binary string is due to chance or whether it can be dismissed as an aberration. Put another way, if you do not know the origins of this string, it is legitimate to ask whether it is likely to be the output of a random process or whether it is a contrived and manipulated succession of digits. Let us see how a statistician can negate the hypothesis that it came from a random mechanism, with a prescribed degree of confidence, by applying de Moivre's theorem. In the computations below n is chosen to be 25, which, as it turns out, is large enough for the normal law to be applied without undue error and small enough to avoid the numbers from becoming unwieldy.

Suppose you are given the following string of 25 zeros and ones, which seems to have a preponderance of ones, and you ask whether it is likely to have been generated from a random process with $p = \frac{1}{2}$; statisticians would call this a *"null hypothesis,"* the straw man of hypothesis testing:

1 1 1 0 1 1 0 0 0 1 1 0 1 1 1 1 1 1 1 0 1 1 1 1 0

I now carry out a computation that is similar to the one done a bit earlier in this chapter. With S_n denoting the number of ones, let us adopt a *confidence level* of .95, meaning that there is a probability .95 that the sample average S_n/n differs from $\frac{1}{2}$ in magnitude by less than $1/\sqrt{n}$. Since $n = 25$, $1/\sqrt{n}$ equals

1/5. The sample average is therefore expected to lie within the interval of ½ plus or minus ⅕, namely (.4, .6), in 19 out of 20 cases. There are actually 18 ones in the given string, and so S_n/n equals 18/25 = .72. This number lies outside the bounds of the confidence interval, and therefore the null hypothesis that p equals ½ is rejected at the .95 confidence level in favor of the alternative hypothesis that p does not equal ½. Note the word *reject*. If the sample average did in fact lie within the confidence interval, we would not be accepting the hypothesis of $p = ½$ but simply not rejecting it. This is because there remains some doubt as to whether the string was randomly generated at all. It could have been deliberately set up with the intent to deceive, or perhaps it represents the outcome of Bernoulli p-trials in which p is not equal to ½. Consider, in fact, the string below generated with $p = .4$:

0 1 0 1 1 1 0 0 1 0 0 1 0 1 1 1 1 1 1 0 0 1 0 0 0

In this case $S_n = 13/25 = .52$, well within the designated confidence interval, even though the hypothetical coin that generated the string is biased. Similarly, though the sequence

0 1 0 1 0 1 0 1 0 1 0 1 0 1 0 1 0 1 0 1 0 1 0 1 0

has a regularly repeating pattern, we find that S_n/n is 12/25 = .48, and therefore the null hypothesis of randomness cannot be rejected in spite of the suspicious nature of the string!

The decision not to reject is a cautionary tactic in light of the evidence, with the agreement that a 95% confidence level is the demarcation line for disbelief. One could equally well adopt the more stringent test of a 99% confidence level, but 95% is often good enough for purposes of reliable inference. As Peter Bernstein comments in his book *Against the Gods,* the situation is analogous to that adopted in a court of law in which criminal defendants do not attempt to prove their innocence but rather their lack of guilt. The prosecutor sets up a null hypothesis of guilt, and the defending attorneys try to reject that assumption on the basis of the available evidence. The accused is found to be guilty or not guilty consistent with the evidence, but innocence is never the issue. The flip side of the coin, however, is that one can err by rejecting a particular sequence, such as a string of mostly zeros, even though it might have, in fact, been generated by a Bernoulli ½-process of uniformly distributed outcomes. Not only can a truly guilty person be set free, but an innocent soul might be judged guilty!

The test invoked above is evidently not a powerful discriminator of randomness. After all, it only checks on the relative balance between zeros and ones as a measure of chance. I included it solely to illustrate the normal law and to provide the first, historically the oldest, and arguably the weakest of tools in the arsenal needed to field the question "Is it random?"

There are a host of other statistical tests, decidedly more robust and revealing, to decide with what Bernoulli called "moral certainty" that a binary sequence is random. One could, for example, check to see whether the total number of successive runs of zeros or ones or the lengths of these runs are consistent with randomness at some appropriate confidence level. A *run* of zeros means an unbroken succession of zeros

What Is Random?

in the string flanked by the digit one or by no digit at all, and a run of ones is defined in a similar manner. Too few runs, or runs of excessive length, are indicators of a lack of randomness, as are too many runs in which zero and one alternate frequently. This can be arbitrated by a statistical procedure called the *runs test* that is too lengthy to be described here. But let us look at two examples to see what is involved. The sequence 00000000000001110000000000 has two runs of zeros and a single run of ones, whereas the sequence 0101010101010101010101010 has twelve runs of one and thirteen of zero. Neither of these sequences would pass muster as random using a runs test at some appropriate confidence level, even though the use of de Moivre's theorem does not reject the randomness of the second string, as you saw.

Examining runs and their lengths raises the interesting question of how people perceive randomness. The tendency is for individuals to reject patterns such as a long run as not typical of randomness and to compensate for this by judging frequent alternations between zeros and ones to be more typical of chance. Experiments by psychologists who ask subjects to either produce or evaluate a succession of digits reveal a bias in favor of more alternations than an acceptably random string can be expected to have; people tend to regard a clumping of digits as a signature pattern of order, when in fact the string is randomly generated. Confronted with HHHTTT and HTTHTH, which do you think is more random? Both, of course, are equally likely outcomes of tossing a fair coin.

Incidentally, people also tend to discern patterns whenever an unequal density of points occurs spatially, even if the points are actually distributed by chance. The emergence of clumps in some portions of space may lead some observers to conclude erroneously that there is a causal mechanism at play. A high incidence of cancer in certain communities, for example, is sometimes viewed as the result of some local environmental condition, when in fact, it may be consistent with a random process.

The psychologists Maya Bar-Hillel and Willem Wagenaar comment that an individual's assessment of randomness in tosses of a fair coin seems to be based on the "equiprobability of the two outcomes together with some irregularity in the order of their appearance; these are expected to be manifest not only in the long run, but even in relatively short segments— as short as six or seven. The flaws in people's judgments of randomness in the large is the price of their insistence on its manifestation in the small." The authors provide an amusing example of this when they quote Linus in the comic strip "Peanuts." Linus is taking a true–false test and decides to foil the examiners by contriving a "random" order of TFFTFT; he then triumphantly exclaims, "If you're smart enough you can pass a true or false test without being smart." Evidently, Linus understands that in order for a short sequence of six or seven T's and F's to be perceived as random, it would be wise not to generate it from a Bernoulli $\frac{1}{2}$-process, since this could easily result in a nonrandom-looking string.

What Is Random?

Another example of biased perception is the belief that some basketball players have a "hot hand" that manifests itself by a run of successful shots. Fans ardently believe that the player's winning streak is not random, but the data may actually be consistent with sequentially *independent* Bernoulli *p*-trials for some probability *p* of success (*p* varies, of course, from player to player). What should one conclude about Joe DiMaggio's 56 game hitting streak in 1941?

The illusion of randomness can also be foisted on an unsuspecting observer in other ways. A deck of cards retains a vestigal "memory" of former hands unless it has been shuffled many times. One mix permutes the ordering of the cards, but the previous arrangement is not totally erased, and you can be deluded into thinking that the pack is now randomly sorted.

More About the Law of Large Numbers

It may be helpful to add some supplementary remarks about the Law of Large Numbers in an attempt to pin down more precisely what is meant by a random binary sequence.

Early in the twentieth century the mathematician Emile Borel established a version of Bernoulli's theorem, the so-called *Strong Law of Large Numbers,* that puts a different spin on what the theorem implies. He showed that with probability one (that is, with one hundred percent certainty) the sample averages S_n/n of a Bernoulli *p*-process will differ from *p* by an arbitrarily small amount *for all sufficiently large n*. This says

The Taming of Chance

something more definite about a typical sequence than does the Law of Large Numbers because S_n/n will be close to p within some specified error and will thereafter remain close for all large n, unlike the result of the Law of Large Numbers, which allows S_n/n to fluctuate considerably in subsequent trials. The Law of Large Numbers is a statement about an individual sequence of Bernoulli p-trials, whereas the Strong Law of Large Numbers makes a claim about all such sequences, with probability one. There is one caveat, however. The sample space of all unlimited Bernoulli trials is no longer finite, and so, in contrast to finite sample spaces, an event of probability one does not necessarily mean that it encompasses the entire ensemble of possibilities without exception, nor does the complementary event of probability zero mean the empty set. There is more here than just hairsplitting, but since it does not play an essential role in what follows, I leave the details to the technical notes for this chapter.[1,2]

We saw that whole numbers can be represented by finite binary strings. It turns out that any number, in particular all numbers between zero and one, can be similarly expressed in terms of a binary string that is usually infinite in length. We accept this fact at present and leave the details to Appendix B.

A number between zero and one was defined to be *normal* by Borel (not to be confused with the term "normal" used earlier in conjunction with de Moivre's theorem) if every digit 0 or 1 in the binary string that represents the number appears with equal frequency as the number of digits grows to infinity,

What Is Random?

and additionally, if the proportions of all possible runs of binary digits of a given length are also equal. In other words, in order for x to be normal not only must 0 and 1 appear with equal frequency in the binary representation of x, but so must 00, 01, 10, and 11 and, similarly, for all 8 triplets 000, 001, ..., 111. The same, moreover, must be true for each of the 2^k k-tuples of any length k. Since successive digits are independent, the probability of any block of length k is the same. The numbers that do not share the property of normality may be shown to form a set of probability zero, though I must remind you of what was pointed out earlier, namely that an event of probability zero does not have to be an empty set of sequences.[1,2] For all intents and purposes, however, an event of zero probability can be safely disregarded.

In the next chapter you will see that any reasonable candidate for a random sequence must, at the very least, define a *normal number*. The Strong Law of Large Numbers for Bernoulli ½-trials says that each binary digit appears with equal frequency with probability one, since the sample average, namely the fraction of ones in n trials, tends to ½. Therefore, our current prototype of a quintessential random process, the Bernoulli ½-trial, generates infinite sequences that, with probability one, are to be found among the normal numbers. That's good, but normality says more. Not only are individual digits equally distributed, but so are all blocks of the same size, and no pattern can be observed in favor of any one block of k digits; each occurs independently and with probability $½^k$.

Although we know that most numbers are normal, providing an explicit example of such a number has proven to be an elusive task, and the only widely quoted example, due to D. Champernowne, is the number whose binary representation is found by taking each digit singly and then in pairs and then in triplets, and so on:

0100011011000001010100 . . .

Where We Stand Now

In this chapter I took the point of view that randomness is a notion that pertains to an ensemble of possibilities, whose uncertainty is a property of the mechanism that generates these outcomes and not a property of any individual sequence. Nevertheless, we are loath to call a string of 100 digits random if zero is repeated 100 times, and so what we need is a way of looking at what an individual run of zeros and ones can tell us about randomness regardless of where it came from. The shift in emphasis consists in looking at the arrangement of a fully developed string by itself and not on what generated it. This is one of the tasks before us in the subsequent chapters. In particular, it will be seen that randomness can be rejected when there are simple rules that largely predict the successive digits in a string. Champernowne's number that was discussed in the previous section will fail to qualify as random, for example.

The arguments used in the remainder of the book are quite different from those employed in the present chapter, in which the normal law for large sample sizes was invoked to test an a priori assumption about the randomness of a string. Although this approach remains part of the core canon of statistical theory, the subsequent chapters will focus less on the randomness of the generating process and more on the patterns that are actually produced. There are a number of surprises in store, and in particular, we will again be deluded into believing that a digit string is random when the mechanism that generated it is not random, and vice versa.

Let us close with a quote from the mathematician Pierre-Simon Laplace, himself an important contributor to the theory of probability in the early part of the ninteenth century, who unwittingly anticipated the need for a new approach:

> "In the game of heads and tails, if head comes up a hundred times in a row, then this appears to us extraordinary, because after dividing the nearly infinite number of combinations that can arise in a hundred throws into regular sequences, as those in which *we observe a rule that is easy to grasp,* and into irregular sequences, the latter are incomparably more numerous."

[2]

Uncertainty and Information

Norman . . . looked at a lot of statistics in his life, searching for patterns in the data. That was something human brains were inherently good at, finding patterns in the visual material. Norman couldn't put his finger on it, but he sensed a pattern here. He said, "I have the feeling it's not random."

Michael Crichton, *Sphere*

What Is Random?

Messages and Information

A half-century ago Claude Shannon, mathematician and innovative engineer at what was then called the Bell Telephone Laboratories, formulated the idea of information content residing in a message, and in a seminal paper of 1948, he established the discipline that became known as information theory. Though its influence is chiefly in communication engineering, information theory has come to play an important role in more recent years in elucidating the meaning of chance.

Shannon imagined a source of symbols that are selected, one at a time, to generate messages that are sent to a recipient. The symbols could, for example, be the ten digits 0, 1, ... , 9, or the first 13 letters of the alphabet, or a selection of 10,000 words from some language, or even entire texts. All that matters for the generation of messages is that there be a palette of choices represented by what are loosely designated as "symbols."

A key property of the source is the uncertainty as to what symbol is actually picked each time. The freedom of choice the source has in selecting one symbol among others is said to be its *information content*.

As this chapter unfolds it will become apparent that maximum information content is synonymous with randomness and that a binary sequence generated from a source consisting of only two symbols 0 and 1 cannot be random if there is some restriction on the freedom the source has in picking either of these digits. The identification of chance with information

Uncertainty and Information

will allow us to quantify the degree of randomness in a string and to answer the question "is it random?" in a different manner from that of the previous chapter.

The simplest case to consider is a source alphabet consisting of just two symbols, 0 and 1, generating messages that are binary strings. If there is an equal choice between the two alternative symbols, we say that the information in this choice is one *bit*. Just think of a switch with two possible positions 0 and 1, in which one or the other is chosen with probability ½. With two independent switches, the number of equally probable outcomes is 00, 01, 10, 11, and it is said that there are two bits of information. With three independent switches there are $2^3 = 8$ possible outcomes, 000, 001, 010, 011, 100, 101, 110, 111, each consisting of three bits of information; in general, n switches result in an equal choice among 2^n possibilities, each coded as a string of n zeros and ones, or n bits.

You can regard the strings as messages independently generated one at a time by a source using only two symbols, or alternatively, the messages may themselves be thought of as symbols each of which represents one of the 2^n equally probable strings of length n. With messages as symbols the source alphabet consists of 2^n messages, and its information content is n bits, whereas a binary source has an information content of one bit.

There is uncertainty as to which message string encapsulated by n binary digits is the one actually chosen from a message source of 2^n possible strings. As n increases, so does the uncertainty, and therefore the information content of the source becomes a measure of the degree of uncertainty as to

What Is Random?

what message is actually selected. Once picked, however, the uncertainty regarding a message is dispelled, since we now know what the message is. Most strings can be expected to have no recognizable order or pattern, but some of them may provide an element of surprise in that we perceive an ordered pattern and not trash. The surprise in uncovering an unexpected and useful string among many increases as the information content of the source gets larger. However, order (and disorder) is in the eye of the beholder, and selection of any other string, patterned or not, is just as likely and just as surprising.

The n bits needed to describe the 2^n messages generated by a binary source are related by the mathematical expression $n = \log m$, where here "log" means "base = 2 logarithm of." The quantity $\log m$ is formally defined as the number for which $2^{\log m} = m$. Since $2^0 = 1$ and $2^1 = 2$, it follows from the definition that $\log 1 = 0$ and $\log 2 = 1$. It is readily seen, moreover, that if $m = 2^n$, then $\log m = n$. Additional properties of logarithms can be found in Appendix C (if you are not familiar with logarithms, just think of them as a notational device to represent certain numerical expressions involving exponents in a compact form). I use logarithms sparingly in this chapter but cannot avoid them entirely because they are needed to formalize Shannon's notion of information content. In fact, the information content of m equally likely and independent choices is defined by Shannon to be precisely $\log m$. With $m = 1$ (no choice) the information is zero, but for $m = 2^n$, the informa-

Uncertainty and Information

tion content is n bits. In the event that a source works from an alphabet of k symbols, all equally likely and chosen independently, it generates a total of $m = k^n$ message strings of length n, each with the same probability of occurring. In this case the information content per string is $\log m = n \log k$, while the information content per symbol is just $\log k$. With three symbols A, B, C, for instance, there are $3^2 = 9$ messages of length 2, namely AA, AB, AC, BA, BB, BC, CA, CB, CC. The information content of the source is therefore twice log 3 or, roughly, 3.17.

The word "information" as used by Shannon has nothing to do with "meaning" in the conventional sense. If the symbols represent messages, a recipient might view one of them as highly significant and another as idle chatter.

The difference between information and meaning may be illustrated by a source consisting of the eight equally likely symbols A, D, E, M, N, O, R, S that generate all "words" having 10 letters. Most of these words are gibberish, but if RANDOMNESS should come into sight out of the ensemble of $2^{10} = 1024$ possible words, there is good reason to be startled and feel that perhaps we are recipients of an omen. Nevertheless, the word RRRRRRRRRR is just as likely to appear and should surprise us no less. Another example would be a source whose symbols consist of a set of instructions in English. The information content of this source tells us nothing about the significance of the individual symbols. One message might instruct you to open a drawer and read the contents, which turn out to be a complete description of the human genome. The unpacked message conveys enormous meaning to a geneticist. The next message directs you to put out the cat, which is arguably not very significant (except, of course, to the cat).

Entropy

After these preliminaries we consider m possibilities each uninfluenced by the others and chosen with perhaps unequal probabilities p_i for the ith symbol, i being any integer from 1 to m. The sample space consists of m elementary events "the ith symbol is chosen." These probabilities *sum to one,* of course, since the m mutually exclusive choices exhaust the sample space of outcomes. The *average information content* of this source, denoted by H, was defined by Shannon to be the negative of the sum of the logarithms of the p_i, each weighted by the likelihood p_i, of the ith symbol:

$$H = -(p_1 \log p_1 + p_2 \log p_2 + \ldots + p_m \log p_m)$$

or, in more compact notation,

$$H = -\text{the sum of } p_i \log p_i.$$

The expression H is called the *entropy* of the source and represents the average information content of the source, in *bits per symbol.* Though this definition seems contrived, it is exactly what is needed to extend the idea of information to symbols that appear with unequal frequencies. Moreover, it re-

Uncertainty and Information

duces to the measure of information content considered previously for the case in which the *m* symbols are all equally likely to be chosen. When the source consists of $m = 2^n$ message strings, each having perhaps different probabilities of occurring, H is regarded as the average information content in *bits per message*.

To illustrate the computation of entropy consider a board divided into 16 squares of the same size and suppose you are asked to determine which square (see Figure 2.1) has some object on it by engaging in a variant of the game "twenty questions." You ask the following questions having yes or no answers:

> Is it one of the 8 squares on the top half of the board? (No)
> Is it one of the 4 squares on the right half of the remaining 8 possibilities? (Yes)
> Is it one of the 2 squares in the top half of the remaining 4 possibilities? (Yes)
> Is it the square to the right of the 2 remaining possibilities? (No)

Letting one mean yes and zero no, the uncovered square is determined by the string 0110 because its location is determined by no yes yes no. Each question progressively narrows the uncertainty, and the amount of information you receive about the unknown position diminishes accordingly. There are $16 = 2^4$ possible squares to choose from initially, all equally likely, and therefore 16 different binary strings. The uncertainty before the first question is asked, namely the entropy, is

What Is Random?

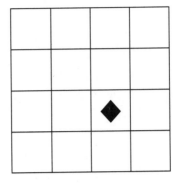

Figure 2.1 A board divided into 16 squares, all of them empty except for one, which contains an object whose location is to be determined by a blindfolded contestant in a variant of the game "twenty questions."

consequently log 16 = 4 bits per string. The entropy decreases as the game continues.

For a binary source consisting of only two symbols with probabilities p and $1-p$, the expression for the entropy H simplifies to

$$H = -[p \log p + (1 - p) \log(1 - p)].$$

Figure 2.2 plots the values of H for a binary source versus the probability p, and we see that H is maximized when $p = \frac{1}{2}$, at which point $H = 1$ (since log 2 = 1); otherwise, $H < 1$. With $p = \frac{3}{4}$, for example, the entropy H is $-\{.75 \log .75 + .25 \log .25\}$, which is roughly .81.

Uncertainty and Information

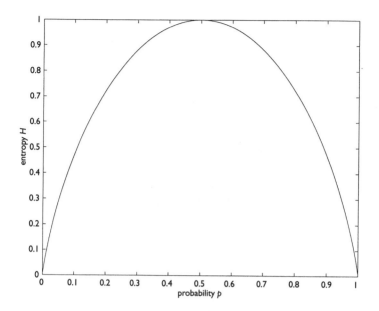

Figure 2.2 Plot of entropy H versus probability p for a binary source consisting of only two symbols with probabilities p and $1 - p$. Note that H is maximized when p is ½ and is zero when there is no uncertainty (p equal to either 0 or 1).

In general, for a nonbinary source, it is possible to show that *H is maximized when the m independent choices are equally likely,* in which case H becomes $\log m$, as already noted. In conformity with the discussion in the preceeding chapter, *maximum entropy is identified with quintessential randomness.* This is illustrated by the board game of Figure 2.1, where the entropy is

What Is Random?

maximum, since all four choices are independent and have an equal probability ½ of being true. The entropy concept will allow us to give new meaning in the next section to the question "Is it random?"

It is important to emphasize that information is equated with uncertainty in the sense that the more unaware we are about some observation or fact, the greater is the information we receive when that fact is revealed. *As uncertainty increases so does the entropy,* since entropy measures information, the resolution of uncertainty. A simple choice among two alternative messages is mildly informative, but having one message confirmed out of many has a degree of unexpectedness and surprise that is very informative. With no choice there is no information, or zero entropy. Message content is irrelevant, however, as we have already stressed. The entropy in a binary source is unaffected by the fact that one symbol represents a thousand-page textbook and the other a simple "duh."

Messages, Codes, and Entropy

If the probabilities p and $1 - p$ in a binary source are unequal, then patterns of repeated zeros or ones will tend to recur, and the ensemble has some redundancy. If it is possible to reduce the redundancy by compressing the message, then a shorter binary sequence can serve as a code to reproduce the whole. In a truly random string all the redundancy has been squeezed out and no further compression is possible. The string is now as short as possible.

Uncertainty and Information

Because of redundancies, a few messages out of the total generated by a binary source are more likely than others due to the fact that certain patterns repeat. There is therefore a high probability that some small subset of messages are actually formed out of the total that are possible. The remaining (large) collection of messages has a small probability of appearing, from which it follows that one can encode most messages with a smaller number of bits.[2.1] Of all the messages of length n that can conceivably be generated by Bernoulli p-trials, the chances are that only a small fraction of them will actually occur when p is not equal to $½$. Instead of requiring n bits to encode each message string, it suffices to use nH bits to represent all but a (large) subset of strings for a message source of entropy H. The required number of bits per symbol is therefore about $nH/n = H$. This compression in message length is effective only when $H < 1$ (p not equal to $1 - p$), since H equals one when $p = ½$.

Stated in a slightly different way, all messages of size n can be divided into two subsets, one of them a small fraction of the total containing messages occurring a large percentage of the time, and a much larger fraction consisting of the messages that rarely appear. This suggests that short codes should be used for the more probable subset and longer codes for the remainder in order to attain a lower average description of all messages. The exception to this, of course, is when $p = ½$, since all messages are then equally probable.

In order to take advantage of message compression, we must find a suitable encoding, but how to do so is less than obvious. I will describe a reasonable, but not maximally efficient,

What Is Random?

coding here and apply it to a specific sequence. Our interest in exploring this now is in anticipation of the discussion in Chapter 4 where I follow Andrei Kolmogorov and others who say that a string is random if its shortest description is obtained by writing it out in its entirety. There is no compressibility. If, however, a shorter string can be found to generate the whole, a shorter string that encodes the longer one, then there must have been recognizable patterns that reduced the uncertainty. The approach taken by Kolmogorov is different than Shannon's, as we will see, since it depends on the use of computable algorithms to generate strings, and this brings in its wake the paraphernalia of Turing machines, a topic that is explained in Chapter 4.

Take a binary string that is created by Bernoulli p-trials in which the chance of a zero is .1 and that of a one is .9 ($p = .9$). Break the string up into consecutive snippets of length 3. Since blocks of consecutive ones are more likely to appear than zeros, assign a 0 to any triplet 111. If there is a single zero in a triplet, code this by a 1 followed by 00, 01, or 10 to indicate whether the zero appeared in the first, second, or third position. In the less frequent case of 2 zeros begin the code with a prefix of 111 followed by 0001 (first and second positions), 0010 (first and third positions) or 0110 (second and third positions). The rare event of all zeros is assigned 11111111. Consider, for example, the string fragment 110 111 101 010 111 110 111 consisting of seven triplets with a total of 21 bits. These triplets are coded, in sequence, by the words 110, 0, 101, 1110010, 0, 110, 0. The code requires 19 bits, a slight

Uncertainty and Information

compression of the original message. In general, the reduction is considerably better, since a much larger number of ones will occur on average for an actual Bernoulli $9/10$-trials than is indicated by the arbitrarily chosen string used here.

Notice that this coding scheme has the virtue of being a *prefix code,* meaning that none of the code words are prefixes of each other. When you come to the end of a code word you know it is the end, since no word appears at the beginning of any other code word. This means that there is a unique correspondence between message triplets and the words that code them, and so it is possible to read the code backwards to decipher the message. This is evident in the example above, in which none of the seven code words reappears as a prefix of any of the other words. You can now confidently read the fragment 11001110010 from left to right and find that 1 by itself means nothing, nor does 11. However, 110 means 110. Continuing, you encounter 0, which denotes 111. Proceeding further, you need to read ahead to the septuple 1110010 to uncover that this is the code for 010; nothing short of the full seven digits has any meaning when scanned from left to right. Altogether, the deciphered message is 1101010.

Although the discussion of codes has been restricted to only two symbols, zero and one, similar results apply to sources having alphabets of any size k. Suppose that the ith symbol occurs with probability p_i with $i = 1, \ldots, k$. Shannon established that there is a binary prefix coding of the k symbols in which the average length of the code words in bits per symbol is nearly equal to the entropy H of the source.[2.2] As a consequence of

What Is Random?

this, a message of length n can be encoded with about nH bits. Consider, for example, a source with four symbols a, b, c, d having probabilities $½, ¼, ⅛$, and $⅛$, respectively. The entropy is computed to be 1.75, and if the code words are chosen as 0 for a, 10 for b, 110 for c, and 111 for d, then this corresponds to code world lengths of 1, 2, 3, 3, whose average length is 1.75, which, in this case at least, equals the entropy. In the example given in the two preceding paragraphs, the source consisted of the eight triplets 000, 001, . . . , 111 of independent binary digits in which the probability of 111, for example, is $9/10$ multiplied by itself three times, or about .73, with a similar computation for all the other triplets. Although a reasonably effective code was devised for this source, it is not as efficient as the code established by Shannon.

In order to attain additional insight into randomness, the assumption that successive symbols are independent can be abandoned by allowing each symbol to be contingent on what the previous symbol or, for that matter, on what several of the previous symbols happened to be. More exactly, the probability that a particular symbol takes on any of the k alphabet values is conditioned on which of these values were assumed by the preceeding symbol(s).

For a simple illustration pick the $k = 2$ case where the alphabet is either yes or no. Suppose that the probability of yes is $⅓$ if the preceeding bit is yes, and $¼$ if the predecessor happened instead to be no. The conditional probabilities of getting no, by contrast, are $⅔$ and $¾$, respectively. This means that if the source emits the symbol yes, then the odds of getting

Uncertainty and Information

no on the next turn are twice that of obtaining yes. There is a built-in redundancy here due to the correlation between successive bits.

A more striking example of sequential correlations is the English language. English text is a concatenation of letters from a total alphabet of 27 (if one counts spaces), or, taking a more liberal viewpoint, a text is a string of words in which the "alphabet" is now a much larger, but still finite, ensemble of possible words.

The frequency with which individual letters appear is not uniform, since E, for example, is more likely than Z. Moreover, serial correlations are quite evident: TH happens often as a pair, for example, and U typically follows Q. To analyze a source that spews forth letters, assume that each letter is produced with different frequencies corresponding to their actual occurrence in the English language. The most frequent letter is E, and the probability of finding E in a sufficiently long text is approximately .126, meaning that its frequency of occurrence is .126, whereas Z has a frequency of only about .001. The most naive simulation of written English is to generate letters according to their empirically obtained frequencies, one after the other, independently. A better approximation to English is obtained if the letters are not chosen independently but if each letter is made to depend on the preceeding letter, but not on the letters before that one. The structure is now specified by giving the frequencies of various letter pairs, such as QU. This is what Shannon called the "digram" probabilities of the paired occurrences. After a letter is chosen, the next one

What Is Random?

is picked in accordance with the frequencies with which the various letters follow the first one. This requires a table of conditional frequencies. The next level of sophistication would involve "trigram" frequencies in which a letter depends on the two that come before it.

Generating words at random using trigram frequencies gives rise to a garbled version of English, but as you move forward with tetragrams and beyond, an approximation to the written word becomes ever more intelligible. Here, for example, is a tetra, or fourth-order approximation, as given by Robert Lucky in his entertaining and informative book *Silicon Dreams:* "the generated job providual better trand the displayed code, abovery upondults well the coderst in thestical it do bock bothe merg. Instates cons eration. Never any of puble and to theory."

The crudest approximation to English is to generate each of the 27 symbols independently with equal probabilities $1/27$. The average entropy per letter in this case is $\log 27 = 4.76$. By the time one reaches tetragram structure, the entropy per symbol is reduced to 4.1, which shows that considerable redundancy has accumulated as a result of correlations in the language induced by grammatical usage and the many habits of sentence structure accrued over time. Further refinements of these ideas led Shannon to believe that the entropy of actual English is roughly one bit per symbol. This high level of redundancy explains why one can follow the gist of a conversation by hearing a few snatches here and there in a noisy room even though some words or entire phrases are drowned out in the din.

Uncertainty and Information

Since redundancy reduces uncertainty, it is deliberately introduced in many communication systems to lessen the impact of noise in the transmission of messages between sender and recipient. This is accomplished by what are called *error-correcting codes,* in which the message, say one block of binary digits, is replaced by a longer string in which the extra digits effectively pinpoint any error that may have occurred in the original block. To illustrate this in the crudest possible setting imagine message blocks of length two designated as $b_1 b_2$ in which each b_i is a binary digit. Let b_3 equal 0 if the block is 00 or 11, and 1 if the block is 01 or 10. Now transmit the message $b_1 b_2 b_3$, which always has an even number of ones. If an error occurs in transmission in which a single digit is altered, as, for example, when 011 is altered to 001, the receiver notes that there is an odd number of ones, a clear indication of an error. With an additional refinement to this code, one can locate exactly at what position the error took place and thereby correct it. Compact disk players that scan digitized disks do an error correction of surface blemishes by employing a more elaborate version of the same idea.

Diametrically opposite to error correction is the deliberate insertion of errors in a message in order to keep unauthorized persons from understanding it. The protection of government secrets, especially in time of warfare, is a remarkable tale that reads like a thriller, especially in David Kahn's *The Code-Breakers.* A secure method of enciphering messages for secrecy, called the Vernam code, after Gilbert Vernam, who devised the first prototype for AT&T in 1917, consists in coding each

What Is Random?

letter of the alphabet by a five-digit binary string. The $2^5 = 32$ possible quintuplets encompass all 26 letters as well as certain additional markers. A message is then some long binary string s. Let v, called the key, designate a random binary string of the same length as s (v is obtained in practice from a "pseudo-random" generator, as described in the next chapter). The cyphered message # is obtained from the plain-text message s by adding the key v to s, digit by digit, according to the following rule:

$$0 + 0 = 0$$
$$0 + 1 = 1$$
$$1 + 0 = 1$$
$$1 + 1 = 0$$

The recipient of # is able to decode the encrypted message by the same procedure: Simply add v to # using the same rule for addition as the one above, and this restores s! For example, if $s = 10010$ and $v = 11011$, the transmitted message # is 01001. The decoded message is obtained from 01001 by adding 11011 to obtain 10010, namely the plain-text s (see Figure 2.3 for a schematic illustration of the entire communication system). Since the key v is random, and known only to sender and receiver, any spy who intercepts the garbled message is foiled from reading it. For reasons of security, the key is usually used only once per message in Vernam's scheme. This to avoid having tell-tale patterns from the analysis of several surreptitious interceptions.

Uncertainty and Information

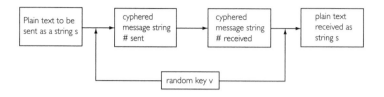

Figure 2.3 Schematic representation of an encipher–decipher coding scheme for sending messages in a secure manner.

There is a balance between the inherent structure and patterns of usage in a language like English and the freedom one has to generate a progression of words that still manages to delight and surprise. A completely ordered language would be predictable and boring. At the other extreme, the random output of the proverbial monkey banging away at a keyboard would be gibberish. English is rich in nuance and comprehensible at the same time. In the last chapter of the book this interplay between chance and necessity becomes a metaphor for the span of human culture and nature at large, in which randomness will be seen as an essential agent of innovation and diversity.

Approximate Entropy

At the end of the previous chapter I stated the intention of deciding the randomness of a given string regardless of its provenance. We are ignorant of how the string came to be, and we don't care. The mechanism that generated it, whether it be pure happenstance or some hidden design, is irrelevant, and we want to distance ourselves from the source and concentrate on the string itself. Taking a cue from the discussion in the preceeding section the question of randomness hinges on the information content of a given binary sequence. A rough and ready tool for assessing randomness in terms of its entropy is called *approximate entropy* (or ApEn, for short).

Suppose that a finite sequence is to some extent by virtue of sequential dependencies, as happens when the likelihood of a zero or one is contingent on whether it is preceded by a one or more zeros or ones. We calculate the degree of redundancy, or, to put it another way, the extent of randomness, by computing an expression analogous to entropy that measures the uncertainty in bits per symbol. If the source generates independent symbols having different probabilities p_i, for $i = 1, 2 \ldots, m,$ then as you have already seen, the entropy is given by the expression $H = -$ the sum of $p_i \log p_i$. However, not knowing the source requires that these probabilities be estimated from the particular sequence before us. The Law of Large Numbers tells us that the p_i are roughly equal to the fraction of times that the ith symbol appears in the given string.

Coming next to *digrams,* namely pairs of consecutive symbols, an expression for the "entropy" per couple is defined in the same way, except that p_i now refers to the probability of obtaining one of the $r = m^2$ possible digrams (where m is the number of symbols and r is the number of digrams formed from these symbols); the expression for H remains the same,

Uncertainty and Information

but you now must sum over the range $i = 1, 2, \ldots, r$. For example, if $m = 3$ with an alphabet consisting of A, B, C, then there are $3^2 = 9$ digrams:

AA	BA	CA
AB	BB	CB
AC	BC	CC

Triplets of consecutive symbols, or *trigrams,* are handled in an analogous manner, allowing for the fact that there are now m^3 possibilities. The same principle applies to k-grams, blocks of k symbols, for any k. What I am striving to establish is that a finite string is random if its "entropy" is as large as possible and if digrams, trigrams, and so forth, provide no new clue that can be used to reduce the information content.

For a given string of finite length n it is necessary to estimate the probabilities p_i for all possible blocks. As before, the Law of Large Numbers assures us that when n is sufficiently large, the value of p_i is roughly equal to the proportion of times that the ith block of length k appears (blocks of length 1 refers to one of the m individual symbols), where i ranges from 1 to m^k. It can be readily determined that there are exactly $n + 1 - k$ blocks of length k in the string, namely the blocks beginning at the first, second, $\ldots$, $(n + 1 - k)$th position. For example, if $n = 7$ and $k = 3$, the blocks of length k in the string ENTROPY are ENT, NTR, TRO, ROP, and OPY, and there are $n + 1 - k = 7 + 1 - 3 = 5$ of them.

Let n_i indicate the number of times that the ith block type occurs among the $n + 1 - k$ successive k-grams in the string.

What Is Random?

Then the probability p_i of the ith block is estimated as the frequency n_i/N, in which we agree to write $(n + 1 - k)$ simply as N.

For example, the string CAAABBCBABBCABAACBACC has length $n = 21$, and m equals 3. There are $m^2 = 9$ possible digrams, and since $k = 2$, these are found among the $N = 20$ consecutive blocks of length 2 in the string. One of the 9 conceivable digrams is AB and this occurs three times, and so the probability of this particular block is 3/20.

It is not hard to see that the n_i must sum to N as i ranges from 1 to m^k. The following approximate expression is then obtained for the "entropy" $H(k)$ per block of size k:

$$H(k) = - \text{ the sum of } (n_i/N) \log (n_i/N),$$

where the index i goes from 1 to m^k; $H(1)$ is identical to the usual expression for H. You should be aware, however, that "entropy" is an abuse of language here, since entropy as defined earlier for H applies to an alphabet of symbols chosen independently, whereas the blocks of size k may be correlated due to sequential dependencies. Moreover, entropy assumes there is some specific generating source, while now there is only a single string from some unknown source, and all we can do is estimate the source from the roughly estimated probabilities.

I can now introduce the key notion of *approximate entropy* ApEn(k) as the difference $H(k) - H(k-1)$, with ApEn(1) being simply $H(1)$. The idea here is that we want to estimate the "entropy" of a block of length k conditional on knowing its prefix of length $k-1$. This gives the *new information con-*

Uncertainty and Information

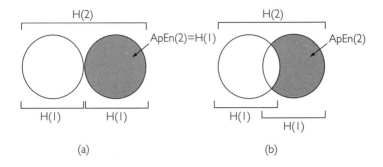

Figure 2.4 Schematic representation of two successive symbols that are sequentially independent (Fig. 3.4a) and dependent (Fig. 3.4b). The degree of overlap indicates the extent of sequential dependence. No overlap means independence. The average "entropy" per disk (read: symbol) is $H(1)$, and the average "entropy" of the digram is $H(2)$, while the shaded area represents the information contributed by the second member of the digram, which is not already contained in its predecessor, conditional on knowing the first member of the pair, namely $H(2) - H(1) = \text{ApEn}(2)$.

tributed by the last member of a block given that we know its predecessors within the block. If the string is highly redundant, you expect that the knowledge of a given block already largely determines the succeeding symbol, and so very little new information accrues. In this situation the difference ApEn will be small. This can be illustrated for digrams schematically in Figure 2.4, in which the ovals represent the average entropy of a pair of

What Is Random?

overlapping (sequentially dependent) symbols in one case and disjoint (independent) symbols in the other. Individual ovals have "entropies" $H(1)$, while their union has "entropy" $H(2)$. The shaded area in diagram (b) represents the new information contributed by the second member of the digram that is not already contained in its predecessor. This is the approximate entropy ApEn(2) of the digram, conditional on knowing the first member of the pair. The shaded area in diagram (a), and therefore the uncertainty, is maximal whenever the ovals do not overlap.

Another way of seeing why ApEn(2) is close to zero when the second digit in a digram pair is completely determined by the preceeding one is that there are about as many distinct digrams in this situation as there are distinct digits, and so $H(1)$ and $H(2)$ are about equal; hence their difference is close to zero. By contrast, if a string of length n is random, then $H(1) = \log n$ and $H(2) = \log n^2 = 2 \log n$, since there are n^2 equally probable digrams; therefore, $\text{ApEn}(2) = 2 \log n - \log n = \log n$, which equals ApEn(1).[2.3]

Again, Is It Random?

For strings that are not too long, it usually suffices to check ApEn(k) for k not exceeding 3 to get an idea of the degree of redundancy. There is a simple algorithm to compute ApEn for binary strings, written as a MATLAB program,[2.4] and I illustrate it here for three different sequences of length 24 with k equal to 1 and 2. The first one repeats the motif of 01, and so

Uncertainty and Information

the patterns 00 and 11 never appear. The second sequence is from Bernoulli ⅕-trials in which some redundancy is expected due to long blocks of consecutive zeros. The last sequence is from Bernoulli ½-trials, and here one may anticipate something closer to maximum "entropy." Since these strings progress from orderly to scrambled, the values of ApEn should reflect this by increasing, and this is precisely what happens. Even though we happen to know the provenance of each series, they could have appeared to us unannounced, and so it is at least conceivable that each could have arisen from a totally ordered or, perhaps, a hopelessly tangled process. We know not which. The sequences of digits are taken as they are, au naturel, without prejudice as to the source, and we ask to what extent the mask they present to us mimics the face of randomness. The approximate entropy is maximized whenever any of the length-k blocks of digits appear with equal frequency, which is the signature requirement of a *normal number* as it was presented in the previous chapter.

It is now clear that in order for an unlimited binary sequence to qualify as random it must indeed be a normal number. Otherwise, some blocks occur more frequently than others, and as in the case of Bernoulli p-processes in which p differs from ½, the unequal distribution of blocks results in redundancies that can be exploited by a more efficient coding. In particular, a random sequence has the property that the limiting frequencies of zeros and ones is the same. I must caution, however, that when you look at a string that is finite in length there is no way of knowing for sure whether this snippet

What Is Random?

comes from a random process or not. The finite string is regarded simply on its own terms. By contrast, the statistical procedures of Chapter 1 tested the null hypothesis that the string is randomly generated, namely that it comes from Bernoulli ½-trials, but an ineluctable element of uncertainty remains even if the null hypothesis is not rejected, since all that it was capable of checking is the frequency of blocks of length one.

Returning to the examples, let the first string consist of 01 repeated twelve times. The computed values of ApEn(1) and ApEn(2) are 1.000 and .001, respectively. Since there is exactly the same number of ones and zeros, ApEn(1) takes on the maximum value of $\log 2 = 1$, but ApEn(2) is nearly zero because the second digit of any digram is completely determined by the first, and there are no surprises.

The next string is

0 0 0 1 0 0 0 1 0 0 0 0 1 0 0 0 0 0 1 0 0 1 0 0

and we find that ApEn(1) = .738 and ApEn(2) = .684. There are many more zeros than ones, which explains why ApEn(1) is less than the theoretical maximum entropy of 1 that would have prevailed if the these digits were equally distributed in frequency. Also, the value of ApEn(2) indicates that the second member of each digram is only partially determined by its predecessor. Zero usually follows a 0, but sometimes there is a surprise; on the other hand, 0 always follows a 1.

The third example, from Bernoulli ½-trials, is

1 1 1 1 0 0 1 0 0 0 1 0 1 1 1 0 1 0 1 1 0 0 1 0

Uncertainty and Information

and here ApEn(1) = .9950, while ApEn(2) = .9777, both close to the maximum entropy of 1. In this example the proportions of ones and zeros are more nearly balanced than they are in the previous example, and this is also the case for each of the diagrams 00, 01, 10, 11.

For purposes of comparison let us reconsider the string in Chapter 1 whose randomness was rejected at a 95% confidence level using de Moivre's theorem. The string consisted of 25 digits:

1 1 1 0 1 1 0 0 0 1 1 0 1 1 1 1 1 1 1 0 1 1 1 1 0

Application of ApEn to this sequence gives ApEn(1) = .8555 and ApEn(2) = .7406, leaving questionable again the null hypothesis of randomness. It is worth noting that though the patterned string of repeated 01 fooled de Moivre, ApEn provides a more stringent test, since ApEn(2) is nearly zero, as you saw (actually, the example in Chapter 1 had 25 digits instead of 24, but the conclusion remains the same).

It is only fair to add that the entropy test for randomness may require the application of ApEn for higher values of k in order to detect any latent regularities. For instance, a sequence in which 0011 is repeated six times has the property that 0 and 1 as well as 00, 01, 10, 11 all appear with equal frequency, and so it fools ApEn(k) into a premature judgment of randomness if k is limited to one or two. It requires ApEn(3) to uncover the pattern. In fact, ApEn(1) is 1.000, ApEn(2) is .9958, but ApEn(3) equals .0018!

What Is Random?

As a footnote, it is worth noting that although a random number must be *normal,* not every normal number is necessarily random. Champernowne's example from Chapter 1 is generated by a simple procedure of writing 0 and 1 followed by all pairs 00 01 10 11, followed by all triplets 000 001 etc, and this can be encoded by a finite string that provides the instructions for carrying out the successive steps. This will be made clear in Chapter 4, but for now it is enough to state that a sufficiently long stretch of this normal number fails to be random precisely because it can be coded by a shorter string.

The Perception of Randomness

In the previous chapter I mentioned that psychologists have concluded that people generally perceive a binary sequence to be random if there are more alterations between zeros and ones than is warranted by chance alone. Sequences produced by a Bernoulli ½-process, for example, occasionally exhibit long runs that run counter to the common intuition that chance will correct the inbalance by more frequent reversals. The coveted "hot hands" that lead to winning streaks in gambling or in sports seem to negate the independence implied by a random process, and observers of long runs of successes or failures often erroneously attribute this to a memory of past events that persists into the present. In effect they succumb to the fallacy that clusters of one digit or the other signal the presence of luck or, as some would have it, the inscrutable finger of fate.

Uncertainty and Information

The psychologists Ruma Falk and Clifford Konold put a new spin on these observations in a study they recently conducted. A number of participants were asked to assess the randomness of binary sequences that were presented to them, either by visual inspection or by being able to reproduce the string from memory. It was found that the perception of randomness was related to the degree of difficulty the subjects experienced as they attempted to make sense of the sequence. To quote the authors, "Judging the degree of randomness is based on a covert act of encoding the sequence. Perceiving randomness may, on this account, be a consequence of a failure to encode," and elsewhere, "The participants tacitly assess the sequence's difficulty of encoding in order to judge its randomness." Of the two sequences

1 1 1 1 1 1 0 0 0 1 1 0 0 0 0 0 0 0 1 1 1,
1 1 0 1 0 1 0 1 0 1 0 0 0 1 1 0 1 0 1 0 1,

the second, with its excess of alternations, is perceived as more random, even though each string departs equally from the number of runs that a truly random string would be expected to have. It is simply more difficult to reproduce the second from memory.

The experimental results are not inconsistent with the idea that a low value of ApEn betrays a patterned sequence whose sequential redundancies can be favorably exploited by employing a suitable code. Actually, the subjects displayed a systematic bias in favor of perceiving randomness in strings with a moderately higher prevalence of alternations than the maximum value of ApEn would indicate.

What Is Random?

Falk and Konold conclude their paper with a comment on "the interconnectedness of seeing the underlying structure (i.e., removing randomness) and hitting upon an efficient encoding. Learning a foreign language is such an instance; forming a scientific hypothesis is often another . . . once a pattern has been recognized, the description of the same phenomenon can be considerably condensed." In a similar vein, the psychologists Daniel Kahneman and Amos Tversky argue that "random-appearing sequences are those whose verbal descriptions are longest." These assertions echo the sentiments of Laplace quoted at the end of the previous chapter, sentiments that resonate loudly in the remainder of this book. In particular, the complexity of a string as an indication of how difficult it is to encode will utimately be recast as a precise measure of randomness in Chapter 4.

To summarize, the idea of entropy was introduced in this chapter as a measure of uncertainty in a message string, and we saw that whatever lack of randomness there is can be exploited by a coding scheme that removes some of the redundancy. But the story is far from complete. Entropy and information will reappear in the next chapter to help provide additional insights into the question "What is random?"

[3]

Janus-Faced Randomness

"That's the effect of living backwards," the Queen said kindly: "it always makes one a little giddy at first . . . but there's one great advantage in it, that one's memory works both ways."

"The other messenger's called Hatta. I must have two, you know—to come and go. One to come and one to go . . . Don't I tell you?" the King repeated impatiently. "I must have two—to fetch and carry. One to fetch and one to carry."

Lewis Carroll, *Through the Looking Glass*

What Is Random?

Is Determinism an Illusion?

The statistician M. Bartlett has introduced a simple step-by-step procedure for generating a random sequence that is so curious it compels us to examine it carefully, since it will bring us to the very core of what makes randomness appear elusive.

Any step-by step-procedure involving a precise set of instructions for what to do next is called an *algorithm*. Bartlett's algorithm produces numbers one after the other using the same basic instruction, and for this reason it is also referred to as an *iteration*. It begins with a "seed" number u_0 between zero and one and then generates a sequence of numbers u_n, for $n = 1, 2, \ldots$, by the rule that u_n is half the sum of the previous value u_{n-1} plus a random binary digit b_n obtained from a Bernoulli ½-process; the successive values of b_n are independent and identically distributed as 0 and 1. Put another way, u_n can take on one of the two possible values $u_{n-1}/2$ or $½ + u_{n-1}/2$, each equally probable.

The connection between two successive values of u_n is illustrated in Figure 3.1, in which the horizontal axis displays the values of u_{n-1}, and the vertical axis gives the two possible values of the successor iterate u_n.

To make further progress with this sequence and to reveal its essential structure it is convenient, indeed necessary, to represent the successive numbers by a binary string. Any number x between zero and one can be represented as an infinite sum

$$x = \frac{a_1}{2} + \frac{a_2}{4} + \frac{a_3}{8} + \frac{a_4}{16} + \ldots$$

Janus-Faced Randomness

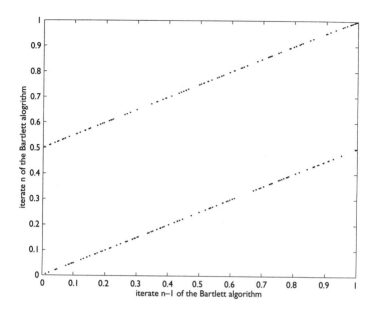

Figure 3.1 Successive values (u_{n-1}, u_n) of the Bartlett algorithm for 200 iterates. For each value of u_{n-1} there are two possible values of u_n, each equiprobable.

in which the coefficients $a_1, a_2, a_3, \ldots$ can be either zero or one. Why this is so is discussed in detail in Appendix B, but for now it is enough to give a few examples. The number $11/16$, for instance, can be expressed as the finite sum $1/2 + 1/8 + 1/16$, while the number $5/6$ is the unending sum $1/2 + 1/4 + 1/16 + 1/64 + 1/256 + \ldots$. The coefficients $a_1, a_2, \ldots$ in the infinite sum can be strung out to form a binary sequence, and this is the representation that I

What Is Random?

am looking for. In the case of $^{11}/_{16}$ the binary sequence is represented as 1011000... with all the remaining digits being 0 (since $a_1 = 1, a_2 = 0, a_3 = 1, a_4 = 1$, and so on), while $^5/_6$ is represented as 11010101..., in which the 01 pattern repeats indefinitely. In the same manner, any number x in the *unit interval* (meaning the numbers between zero and one) can be represented as a binary sequence $a_1 a_2 a_3 \ldots$; this applies, in particular, to the initial value u_0.

The trick now is to see how the first iterate u_1 of Bartlett's algorithm is obtained from u_0 in terms of the binary representation. The answer is that the sequence $a_1 a_2 \ldots$ is shifted one place to the *right* and a digit b_1 is added to the left.[3.1] In effect, u_1 is represented by $b_1 a_1 a_2 a_3 \ldots$, and mutatis mutandis, by shifting the sequence to the right n places and adding random digits on the left, the nth iterate is expressed as $b_n b_{n-1} \ldots b_1 a_1 a_2 a_3 \ldots$. A simple example shows how this works: Pick the initial seed number u_0 to be $¼$ and suppose that the random digits b_1 and b_2 are respectively 0 and 1. Then $u_1 = ⅛$ and $u_2 = ^9/_{16} = ½ + ^1/_{16}$. The binary representations of u_0, u_1, and u_2 are now found to be 01000..., 001000..., 1001000..., in accord with what was just described.

The clever scheme of replacing the iterations of u_n by the easier and more transparent action of shifting the binary sequence that represents u_n is known as *symbolic dynamics*. Figure 3.2 is a schematic representation.

At this point we should pause and note a most curious fact: Bartlett's randomly generated sequence is a set of numbers that,

Janus-Faced Randomness

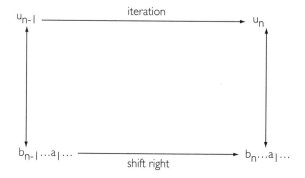

Figure 3.2 Representation of Bartlett's algorithm using symbolic dynamics: Iteration from u_{n-1} to u_n is tantamount to a shift of the binary sequence $b_{n-1} \ldots a_1 \ldots$ to the right and then adding the digit b_n to the left.

when *viewed in reverse,* is revealed as a deterministic process! What I mean is that if you start with the last known value of u_n and compute u_{n-1} in terms of u_n and then u_{n-2} in terms of u_{n-1}, and so on, there is an unequivocal way of describing how the steps of Bartlett's iteration can be traced backward. The iterates in reverse are obtained by shifting the binary sequence that represents u_n to the *left* one step at a time and truncating the leftmost digit: The sequence $b_n b_{n-1} \ldots b_1 a_1 a_2 \ldots$ therefore becomes $b_{n-1} b_{n-2} \ldots b_1 a_1 a_2 \ldots$. This is shown schematically in Figure 3.3.

This shifting operation does not entail the creation of random digits as it does in Bartlett's iteration, but instead, it simply

What Is Random?

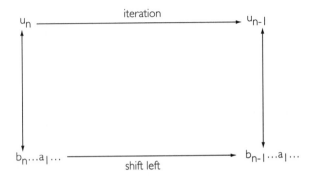

Figure 3.3   Representation of the inverse of the Bartlett algorithm using symbolic dynamics: Iteration from u_n to u_{n-1} is tantamount to a shift of the binary sequence $b_n \ldots a_1 \ldots$ to the left and then deleting the leftmost digit.

deletes existing digits, and this is a perfectly mechanical procedure.

We can now step back a moment to see what this inverse operation actually amounts to. Pick a new seed number v_0 in the unit interval and now generate a sequence v_n for $n = 1, 2, \ldots$ by the iterative rule that v_n is the *fractional part of twice* v_{n-1}. Taking the fractional part of a number is often expressed by writing "mod 1," and so v_n can also be written as $2v_{n-1}$ (mod 1). For instance, if v_{n-1} equals ⅞, then v_n is obtained by doubling ⅞; which yields $^{14}/_8 = 1¾$, and taking the fractional part, namely ¾. Now suppose that v_0 has, like u_0 earlier, an infinite binary representation $q_1 q_2 q_3 \ldots$. It turns out that the action of

Janus-Faced Randomness

getting v_n from its predecessor is represented symbolically by lopping off the leftmost digit and then shifting the binary sequence to the left: $q_1q_2q_3 \ldots$ becomes $q_2q_3q_4 \ldots$.^{3.2} This is illustrated by the sequence of three Bartlett iterates given earlier, namely ¼, ⅛, ⁹⁄₁₆. Beginning with ⁹⁄₁₆ as the value v_0 and working backwards by the "mod 1" rule, you get $v_1 = {}^{18}\!/_{16}(\text{mod } 1) = ⅛$ and $v_2 = ¼(\text{mod } 1) = ¼$. In terms of binary representations the sequence of iterates is $1001000\ldots$, $001000\ldots$, $01000\ldots$.

It is now apparent that the mod 1 iteration is identical in form to the inverse of the Bartlett iteration, and you can again ascertain that this inverse is deterministic, since it computes iterates by a precise and unambiguous rule.

Figure 3.4 plots the relation of v_n to v_{n-1}, and by tilting your head it becomes clear that this picture is the same as that in Figure 3.1 viewed in reverse. All you need do is to relabel v_n, v_{n-1} as u_{n-1}, u_n in that order, as shown in Figure 3.5.

So now we have the following situation: Whenever Bartlett randomly generates a binary string $b_n \ldots b_1 a_1 \ldots$, the mod 1 algorithm reverses this by deterministically deleting a digit to obtain the sequence $b_{n-1} \ldots b_1 a_1 \ldots$. In effect, the future unfolds through Bartlett by waiting for each new zero or one event to happen at random, while the inverse forgets the present and retraces its steps. The uncertain future and a clear knowledge of the past are two faces of the same coin. This is why I call the Bartlett iterates a *Janus sequence,* named after the Roman divinity who was depicted with two heads looking in opposite directions, at once peering into the future and scanning

What Is Random?

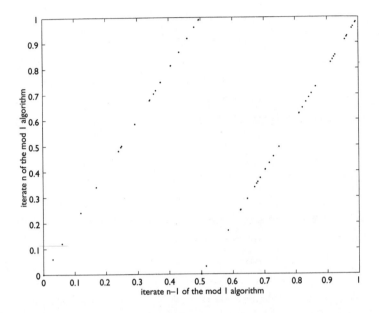

Figure 3.4 Successive values (v_n, v_{n-1}) of the inverse Bartlett algorithm, namely the mod 1 algorithm, for 200 iterates. For each value of v_n there is a unique value of v_{n-1}, a deterministic relationship.

the past. Figure 3.6 shows a particular sequence of values iterated by Janus and its inverse; they are manifestly the same.

Although this Janus-faced sequence appears to be a blend of randomness and determinism, a more careful scrutiny reveals that what poses as order is actually disorder in disguise. In fact, the string in reverse generates what has come to be known as *deterministic chaos,* in which any uncertainty in initial

Janus-Faced Randomness

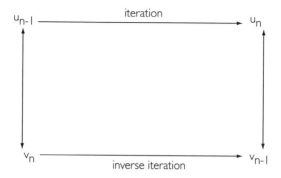

Figure 3.5 The pair (v_n, v_{n-1}) of Figure 3.4 corresponds to the pair (u_{n-1}, u_n) of Bartlett's algorithm (Figure 3.1), whose iterates constitute the *Janus sequence*. The iterates defined by the inverse Bartlett algorithm, namely the mod 1 algorithm, define the *inverse Janus sequence*.

conditions inevitably translates into eventual randomness. In order to see this, let us return to the way the mod 1 iterates shift a binary sequence $q_1 q_2 \ldots$ to the left. If the string $q_1 q_2 \ldots$ corresponding to the initial condition v_0 happens to be known in its entirety, then there is no uncertainty about the number obtained after each shift—it is the number that masquerades as the sequence $q_2 q_3 \ldots$.

The problem arises from what is called *coarse graining*, namely, when v_0 is known only to limited precision as a *finite string* $q_1 \ldots q_n$. The reason is that all infinite binary strings can be regarded as outputs of an unending Bernoulli ½-process

What Is Random?

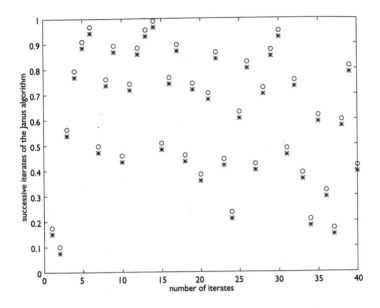

Figure 3.6 Forty Janus iterates (circles) starting from 0.15 compared to forty iterates of the inverse Janus sequence (stars) starting from the last computed value of the Janus iterations plotted in reverse. For ease of identification, a small constant was added to each of the inverse iterates in order to distinguish the backward and forward values (otherwise, they would overlap).

(Appendix B elaborates on this correspondence, as does the technical note 1.2), and therefore, if v_0 is known only to the finite precision $q_1 \ldots q_n$ the remaining digits of the sequence comprise the undetermined remnants of a Bernoulli ½-process. After n iterates the first n digits have been truncated, and the

Janus-Faced Randomness

string now looks like $q_{n+1}q_{n+2}\ldots$, and this consists of a sequence of zeros and ones about which all we know is that they are *independent and equally probable;* from here on out the mod 1 iterates are unpredictable. If q_{n+1} is zero, then the $(n+1)$st iterate is a number in $[0, \frac{1}{2})$, whereas if q_{n+1} equals one, this tells you that the iterate lies somewhere within $[\frac{1}{2}, 1]$; each event has the same probability of occurring.

The seemingly deterministic behavior of the reverse Janus iterates (mod 1 iteration) is evidently compromised unless one knows the present value v_0 exactly as an infinite sequence $q_1 q_2 \ldots$. Moreover, a subtle change in the initial conditions, a simple flip from 0 to 1 in one of the digits, can result in a totally different sequence of iterates (Figure 3.7). Dynamical theorists call the mod 1 algorithm, namely the reverse of Janus, *chaotic* because of this sensitivity to initial conditions: A small change is magnified by successive iterates into unpredictability; *ignorance now spawns randomness later.*

One can produce a random sequence of iterates r_n, $n = 1, 2, \ldots$, using the mod 1 algorithm directly. Starting with a v_0 whose binary expansion corresponds to some random Bernoulli sequence, set r_n to 0 if the nth iterate is a number in $[0, \frac{1}{2})$ and set it equal to 1 if the number is within $[\frac{1}{2}, 1]$. A little reflection shows that even though the successive values of r_n are generated by a purely mechanical procedure, they are identical to the digits q_n given earlier. The iterates thus define a random process. This is a startling revelation because it suggests that determinism is an illusion. What is going on here?

What Is Random?

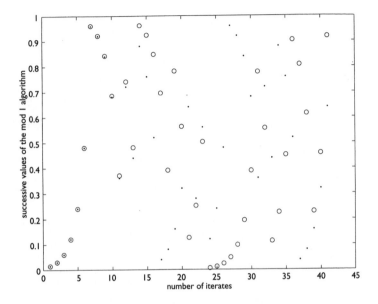

Figure 3.7 Forty iterates of the deterministic inverse Janus algorithm starting at slightly different values of 0.01500 (dots) and 0.01501 (circles) to illustrate sensitivity to initial conditions. For the first few iterates, the dots reside within the circles, but soon they begin to diverge due to the magnification of the initial difference in values, which makes itself felt more and more with successive iterates.

The solution to the apparent paradox has already been given: If v_0 is known completely in the sense of knowing all its binary digits q_n, then the "random" process generated by the mod 1 algorithm collapses into a futile enumeration of an unending and predestined supply of zeros and ones, and determinism is preserved. The problem, as I said earlier, is to truly possess the digits of v_0's binary representation in their entirety, and as a practical matter, this eludes us; the curse of coarse graining is that randomness cannot be averted if v_0 is imperfectly known. On a more positive note, however, it may take many iterates before the unpredictability manifests itself, long enough to provide some temporary semblance of order.

A clarifying comment is needed here. The randomness of the sequence r_n generated above hinges on having v_0 be represented by a Bernoulli ½-trials that is itself random. The Strong Law of Large Numbers assures us that this is true with probability one, as we discussed in Chapter 1, but "probability one" does not rule out certain exceptions, such as sequences that terminate with an unending supply of either all zeros or all ones, or that have a pattern that repeats indefinitely. Nevertheless, these anomalous starting values for v_0 constitute a negligible subset of the totality of binary sequences that are possible and can be safely disregarded.[1,2]

Generating Randomness

The deterministic nature of mod 1 is unmasked in other ways, the most convincing of which is the plot in Figure 3.4, which shows that successive iterates follow each other in an orderly pattern, a consequence of the fact that v_n is dependent on v_{n-1}. By contrast, the successive values of a random sequence are uncorrelated and appear as a cloud of points that scatter haphazardly (Figure 3.8).

What Is Random?

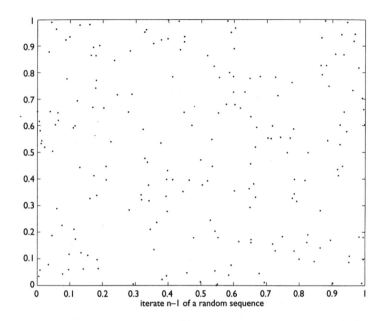

Figure 3.8 Two hundred successive values of a random sequence (obtained by the "pseudo-random generator" to be discussed presently) plotted as pairs (x_{n-1}, x_n). A featureless cloud of points emerges.

I obtained the points in Figure 3.7 by using what is known as a *random number generator* that comes bundled today with many software packages intended for personal computers. They supply "random" numbers by an algorithm not much different from the one employed by the mod 1 procedure: An initial seed is provided as some *integer* x_0, and then integers x_n

Janus-Faced Randomness

for $n = 1, 2, \ldots$ are produced by the deterministic iterative scheme

$$x_n = (a\, x_{n-1} + c) \bmod m.$$

The numbers a, c, and m are positive integers that need to be specified, and their choice moderates the character of the successive iterates. For example, if $a = 2$, $c = 1$, and $m = 8$, then with x_0 chosen as 5, the first iterate x_1 becomes 3.

Since the generated integers don't exceed m, it is fairly evident that at most m distinct integer values of x_n can be produced, and so the iterations must eventually return to some previous number. From here on the procedure repeats itself. This seemingly innocuous comment deserves some elaboration. Computers store numbers in binary form, and the integers from 0 to $m - 1$ are each represented by one of $m = 2^k$ binary strings. This takes place by writing each integer b as a sum of powers of 2. This is explained more fully in Appendix B, but a few examples here will make clear what is involved. Choose, for simplicity, the case $k = 3$. There are then eight binary strings of length 3, and each integer b from 0 to 7 is represented by a binary triplet $a_0 a_1 a_2$ by means of $b = 4a_2 + 2a_1 + a_0$. For instance, 5 is coded by 101 and 7 by 111 because $5 = 4 + 1$ in the first case and $7 = 4 + 2 + 1$ in the second. Similarly, all strings of length 5 represent the integers ranging from 0 to 31, since $2^5 = 32$. One example suffices: $23 = 16 + 4 + 2 + 1$, and so 23 is coded by 10111.

Each integer iterate x_n produced by the random number generator is converted to a fraction within the unit interval by

What Is Random?

dividing x_n by m. The considerations of the previous paragraph imply that the numbers x_n/m, which lie between 0 and 1, are represented by finite-length binary strings, and these may be thought of as truncations of an infinite string whose remaining digits are unknown. This coarse graining, a necessary shortcoming of the limited span of computer memory, is what gives the algorithm its cyclical nature, since it is bound to return to a previous value sooner or later. The idea is to obtain a supply of digits that don't cycle back on themselves for a long time, and at the very least this requires that m be large enough. In many versions of the algorithm m is set to $2^{31} - 1$, indeed quite large, and if a and c are chosen judiciously, every integer between 0 and $m - 1$ will be obtained once before recycling.

The recurrence of digits shows that the iterates are less than random. In spite of this inherent flaw, the sequence $\{x_n\}$ typically manages to pass a battery of tests for randomness, such as the runs test discussed in Chapter 1. The numbers thus qualify as *pseudo-random,* satisfactory enough for most applications. Many users are lulled into accepting the unpredictability of these numbers and simply invoke an instruction like "$x =$ rand" in some computer code whenever they need to simulate the workings of chance.

A primitive example of a *pseudo-random* generator is obtained by letting $a = 2$ and $c = 1$, with $k = 3$. Since m equals 8, the only numbers that can be computed are 0 through 7. With a starting seed value of 2, the succeeding values become 5, 3, 7, 7, . . . and the procedure bogs down in a rut with a cycle of length 1. This example is admittedly crude, but it illustrates

Janus-Faced Randomness

the pitfall of premature cycling, something that a good random number generator is designed to avoid.

Even more sophisticated random number generators betray their inherent determinism, however. If one plots successively generated values as pairs or triplets in a plane or in space, orderly spatial patterns appear as striations or layers (the cloud of points you see in Figure 3.8 is misleading because the delicate crystalline tracery is hard to discern). "Random numbers fall mainly in the planes" is the way mathematician George Marsaglia once put it, to the strains, no doubt, of a tune from *My Fair Lady*.[3.5]

Just how random, incidentally, is the Janus algorithm? Though the successive iterates are generated by a chance mechanism, they are also sequentially correlated. This implies a level of redundancy that the ApEn test of the previous chapter should be able to spotlight. Twenty distinct iterates of the Janus algorithm starting with $u_0 = .15$ were used to plot Figure 3.6, and the application of ApEn(1) to this string gives 4.322, which is simply the entropy of a source alphabet of twenty equally likely symbols ($\log 20 = 4.322$). However, ApEn(2) is merely .074, another clear indication that substantial patterns exist among the iterates.

A truly random sequence of Bernoulli ½-trials consists of equiprobable blocks of binary digits of a given size. This would also have to be true in reverse, since a simple interchange between 0 and 1 results in mirror images of the same blocks. Therefore, any test for randomness in one direction would have to exhibit randomness when viewed backwards.

What Is Random?

Janus and the Demons

The "demons" in question are hypothetical little creatures invoked in the last century to grapple with paradoxes that emerged during the early years of thermodynamics.

The nineteenth-century dilemma was that an ensemble of microscopic particles moving deterministically in some confined vessel according to Newton's laws of motion needed to be reconciled with Ludwig Boltzmann's idea that the entire configuration of particles moves, on the average, from highly ordered to disordered states. This inexorable movement toward disorder introduces an element of irreversibility that seemingly contradicts the reversible motion of individual particles, and this observation was at first seen as a paradox. In the heated (no pun intended) debate that ensured the physicist James Clerk Maxwell introduced a little demon in 1871 who allegedly could thwart and thereby dispel the irreversibility paradox.

The demon, whom Maxwell referred to as a "very observant and nimble-fingered fellow," operates a small doorway in a partition that separates a gas in thermal motion into two chambers. The tiny creature is able to follow the motion of individual molecules in the gas and permits only fast molecules to enter one chamber, while allowing only slow ones to leave. By sorting the molecules in this manner a temperature difference is created between the two portions of the container, and, assuming that initially the temperature was uniform, order is created from disorder (by this I mean that a temperature difference can be exploited to generate organized motion, whereas the helter-skelter motion of molecules at uniform temperature cannot be harnessed in a useful manner). In 1929 the physicist Leo Szilard refined this fictional character and made a connection to the idea of information and entropy, which, of course, dovetails with one of the themes in this book.

Before seeing how Szilard's demon operates, we need to backtrack a little. Classical physics tells us that each molecule in the container has its future motion completely determined by knowing exactly its present position and velocity. For simplicity I confine myself to position only, and divide up the space within the box containing the billions of molecules of the gas into a bunch of smaller cells of equal volume, and agree to identify all molecules within the same cell. In this manner a "coarse graining" is established in which individual particles can no longer be distinguished from one another except to say that they belong to one of a finite number N of smaller cells within the container.

As the molecules bounce off of each other and collide with the walls of the box they move from one cell to another, and any improbable initial configuration of molecules, such as having all of them confined to just one corner of the enclosure, will over time tend to move to a more likely arrangement in which the molecules are dispersed throughout the box in a disorderly fashion. Although it is remotely conceivable that just the opposite motion takes place, Boltzmann established, as mentioned earlier, that on average the motion is always from an orderly to less orderly arrangement of molecules. This is one version of the *Second Law of Thermodynamics,* and it is supported by the

What Is Random?

observation that an ancient temple neglected in the jungle will, over time, crumble into ruins, a heap of stones.

The probability p_i of finding a molecule in the ith cell at some particular time is, for $i = 1, 2, \ldots, N$, very nearly n_i/T according to the Law of Large Numbers, for N large enough, where n_i is the number of molecules actually located in that cell and T is the total number of molecules within all N cells. In the previous chapter we found that the entropy H per cell is the negative of the sum of $p_i \log p_i$. What Boltzmann established, in effect, is that H can never be expected to decrease, and its maximum is attained when the molecules are equidistributed among the cells ($p_i = 1/N$ for all i); total disorder is identified with maximum entropy and randomness.

The exact location of each molecule at some particular time is specified by its three position coordinates, and these numbers can be represented by infinite binary strings, as we've seen. However, the coarse graining of molecules into N cells means that the exact position is now unknown, and the location of a molecule is characterized instead by a binary string of finite length that labels the particular cell it happens to find itself in, much the same as in the game of "twenty questions" of Chapter 2 where an object that is put on a board divided into 16 squares is determined by 4 binary digits. If all the molecules are confined to the bottom-left cell of the 16 cells of Figure 2.1, for example, then each molecule has the same binary label 0000 using the yes–no questioning used in that figure. As the number of cells increase the binary strings that label the positions become longer and the imprecision regarding a molecule's po-

Janus-Faced Randomness

sition diminishes, but as long as there are only a finite number of cells, the whereabouts of each molecule remains imperfect. It is this imprecision that undermines the determinism; there is no uncertainty without coarse graining. To repeat an earlier aphorism: ignorance now spawns randomness later!

An analogy with the random permutation of binary strings may be helpful here. Suppose one is initially given a string of n zeros out of an ensemble of $N = 2^n$ possible binary strings of length n. This unlikely arrangement corresponds to an improbable configuration of all molecules confined to one cell, all other cells being unoccupied. Now choose at random from a selection of strings corresponding to a cluster of cells adjacent to the initial cell. Repeat this process with ever larger cell clusters to mimic the haphazard dispersion of molecules, as time evolves, from an initial tight arrangement to a wider set of possibilities in which more binary digits get altered by chance.

Eventually, the ensemble consists of all N cells, and you expect to find that the string has an equally probable distribution of 0's and 1's; it is very unlikely to find a string that is highly patterned such as, for example, all zeros or a repetition of a block such as 011. The reason for this was explained in Chapter 2, where it was shown, using the Law of Large Numbers, that it is very likely for long binary strings to have a nearly equal distribution of 0' and 1's.

Essentially it comes to this: There are many more ways of getting strings with an equal number of 0's and 1's than there are ways to obtain a string of all zeros. With n equal to 4, for example, there is a single string 0000, but six strings are found in

What Is Random?

which 0 and 1 are in the same proportion, namely 0011, 0101, 0110, 1001, 1010, 1100. As n increases, the discrepancy between ordered (patterned) and disordered (random) strings grows enormously. This corresponds to starting with an implausible arrangement of molecules and finding that, over time, it disintegrates into a more likely arrangement in which the molecules are nearly equidistributed among the N cells of the enclosure. This "one-way street" is the paradox of irreversibility.

However, if the molecules begin naturally in a more probable configuration, then irreversibility is less apparent; indeed, it is possible to observe a momentary move toward increased order! It is the artifice of starting with a contrived (human made) and highly unusual configuration that creates the mischief. In the normal course of unmanipulated events it is the most probable arrangements that one expects to see. Humpty Dumpty sitting precariously on a wall is a rare sight; once broken into many fragments there is little likelihood of his coming together again without some intervention, and even all the King's men may not be able to help.

This brief discussion of irreversibility provides another illustration that a deterministically generated sequence can behave randomly because of our ignorance of all but a finite number of binary digits of an initial seed number in initial conditions. In the present case this is caused by truncating the binary sequences that describe the exact position of each molecule, herding all of them into a finite number of cells.

Let's now return to Szilard and his demon and finally make a connection to the Janus algorithm. A single molecule

Janus-Faced Randomness

is put in a container whose two ends are blocked by pistons and that features a thin removable partition in the center (Figure 3.9). Initially, the demon observes which side of the partition the molecule is on and records this with a binary digit, 0 for the left side and 1 for the right. The piston on the side not containing the molecule is gently pushed toward the partition, which is then removed. This can be accomplished without an expenditure of energy, since the chamber through which the piston is compressed is empty. Moreover, Szilard reasoned that moving the partition can be done with negligible work. The energetic molecule now impinges on the piston and moves it back to where it was. Thermal motion creates useful work, a reversal of the usual degradation of organized motion into the disordered motion of heat, and a seeming violation of the Second Law of Thermodynamics. Szilard figured that this decrease in entropy must be compensated by an equivalent increase in entropy due to the act of measurement by the demon to decide which side of the partition the molecule is on. The contemporary view is that what raises the entropy is the act of erasing the measurement after the piston has returned to its original position. Consider this: When one bit is erased, the information, and therefore the entropy, is increased, since there is now uncertainty as to whether the digit was 0 or 1 prior to erasure.

If the demon does not erase after each movement of the piston, the entropy certainly decreases, but one is left with a clutter of recorded digits that appears as a disordered binary string of length n. To return the system to its original state requires that

What Is Random?

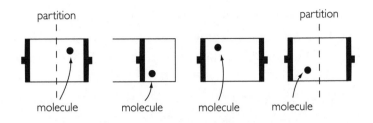

Figure 3.9 The Szilard engine. A demon observes a molecule in the chamber on the right and gently pushes the piston on the left toward the center partition, which is then removed. The molecule in motion impinges on the piston, sliding it back to the left. This creates useful work and lowers the entropy. The cycle then begins anew.

this string be blotted out, and in so doing, n bits are lost, which is tantamount to an increase of uncertainty measured by an entropy of $\log n$. In this manner the Second Law of Thermodynamics maintains its integrity, and it reveals why Maxwell's "little intelligence" cannot operate with impunity. In his stimulating book *Fire In The Mind* George Johnson tells of an imaginary vehicle powered by a "Szilard Engine" whose piston is cycled back and forth by the demon using information as a fuel. The exhaust, a stream of binary digits, pollutes the environment and raises the entropy.

The operation of the demon is mimicked by the action of Janus, shifting a binary sequence (representing an initial configuration) to the right and inserting a binary digit to the left. After n iterations a binary string of length n is created to the left

Janus-Faced Randomness

of the initial sequence, and it is read from right to left. Initially, the n digits are unknown and can be ordered in any one of the 2^n possibilities. Each iterate corresponds to a measurement that decreases the uncertainty and lowers the entropy. After k iterations of Janus have taken place, the information content of the string has been reduced to $n - k$ for $k = 1, 2, \ldots, n$ because only 2^{n-k} strings remain to be considered. At the end, the inverse to Janus deletes the string by repeatedly shifting to the left and chopping off the leftmost digit, and this is read from left to right. Uncertainty is now increased due to the ignorance of what was erased. The information content of the string, after k iterations of the inverse have been applied, is k (namely, $\log 2^k$), since 2^k possible strings have been eliminated and are now unknown.

The Janus algorithm has now been given a new twist as a device that reduces randomness through the act of measuring the future, while its inverse increases uncertainty because it discards acquired digits. Put another way, the entropy decreases with each measurement, since it lessens any surprise the future may hold, but in its wake, there is a trail of useless digits that now serve only as a record of what once was. These junk digits increase entropy, since they represent disorder and randomness. Janus hits its stride when it is combined with its inverse. Abolishing the past after recording the future brings it back to the beginning; entropy is preserved, and the Second Law of Thermodynamics is not violated. The demon has been exorcised by reconciling the two faces of ignorance now and disorder later. Figure 3.10 shows the sequence of

What Is Random?

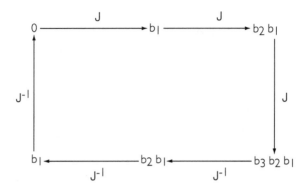

Figure 3.10 The action of the Janus algorithm on the initial value zero (chosen for simplicity of representation), shifting one place to the right. This shift is denoted by J, for Janus, and it accrues a randomly generated digit b_1. Repeated action of J adds digits b_2 and b_3 to the left. The deterministic inverse, denoted by J^{-1} for Janus inverse, shifts one place to the left and chops off the leftmost digit, so that $b_3b_2b_1$ becomes b_2b_1, and then b_2b_1 becomes b_1. J decreases entropy, while J^{-1} increases it. Since J and J^{-1} cancel each other in pairs, the total action is to return to the starting value zero, and entropy is conserved.

steps schematically, starting, for simplicity, with an initial value of zero.

In the next chapter Janus returns to establish a similar reconciliation between the dual views of randomness as information and complexity.

[4]

Algorithms, Information, and Chance

The Library is composed of an . . . infinite number of hexagonal galleries . . . (it) includes all verbal structures, all variations permitted by the twenty-five orthographical symbols, but not a single example of absolute nonsense. It is useless to observe that the best volume of the many hexagons under my administration is entitled The Combed Thunderclap *and another* The Plaster Cramp *and another* Axaxacas mlö. *These phrases, at first glance incoherent, can no doubt be justified in a cryptographical or allegorical manner; such a justification is verbal and, ex hypothesi, already figures in the Library. I cannot combine some characters dhcmrlchtdj*

What Is Random?

> *which the divine library has not forseen and which in one of its secret tongues do not contain a terrible meaning.*
>
> *The certitude that some shelf in some hexagon held precious books and that these precious books were inaccessible seemed almost intolerable. A blasphemous sect suggested that . . . all men should juggle letters and symbols until they constructed, by an improbable gift of chance, these canonical books . . . but the Library is . . . useless, incorruptible, secret.*
>
> Jorge Luis Borges, "The Library of Babel"

Algorithmic Randomness

During the decade of the 1960s several individuals independently arrived at a notion that a binary string is random if its shortest description is the string itself. Among the main protagonists in this story is the celebrated Russian mathematician Andrei Kolmogorov, whom we met earlier, and Gregory Chaitin, information theorist and computer scientist.

The *algorithmic complexity* of a binary string s is formally defined as the length of the *shortest program,* itself written as binary string $s^\star$, that reproduces s when executed on some computer (see Figure 4.1). A "program" in this context is simply an algorithm, a step-by-step procedure, that has been coded into binary form.

Algorithms, Information, and Chance

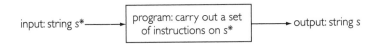

Figure 4.1 Representation of a computer acting on an input string $s^\star$ to produce an output string s.

The idea behind the definition of algorithmic complexity is that some strings can be reproduced from more concise descriptions. For example, if s is a thousandfold repetition of 01 as 010101..., this can be condensed into the statement "copy 01 a thousand times" and, as will be seen, this can be coded by a short binary program $s^\star$ that squeezes out the redundancy. This program requires a fixed number of bits to accommodate the instructions "copy 01 a thousand times" and the number 1000 additionally demands another 10 bits (since 1000 can be written in binary form as 1111101000—see Appendix B). Altogether, the program $s^\star$ is much smaller than s, and when the computer acts on $s^\star$ it produces s as its output. The string s therefore has low algorithmic complexity.

On the other hand, a sequence of zeros and ones generated by a Bernoulli ½-process gives a binary string s that in all likelihood cannot be described by any method shorter than simply writing out all the digits. In this case, the requisite program "copy s" has roughly the same length as s itself, because what one needs to do is to supply the computer with s, and it

What Is Random?

then displays the result. The string has maximum algorithmic complexity.

A string s is said to be algorithmically random if its algorithmic complexity is maximal, comparable to the length of the string, meaning that it cannot be compressed by employing a more concise description other than writing out *s* in its entirety. The code that replicates the string is the string itself and, *a fortiori,* the string must replicate the code.

If a long string *s* of length *n* has a fraction *p* of ones (and therefore a fraction $1-p$ of zeros), then, recalling the arguments in Chapter Two, there is a binary string $s^\star$ of length roughly nH that replicates *s* by exploiting the redundancies in the string. The string $s^\star$ can be constructed as a Shannon code: If *p* is not equal to ½ then the entropy *H* of the source is less than one, and $s^\star$ is shorter than *s*. In order to place this coding within the framework of algorithmic complexity, all that is needed is a program that performs the decoding of $s^\star$. As you may recall, this decoding can be done uniquely in a step-by-step manner, since Shannon's procedure is a prefix code, meaning that none of the code words are prefixes of each other. The string *s* has moderate algorithmic complexity. Figure 4.2 illustrates strings of low, moderate, and maximum complexity.

With $p = 9/10$, for instance, the entropy *H* is about 0.47, and the algorithmic complexity of *s,* being the size of the shortest program, does not exceed 0.47 *n*. From this it is evident that for *s* to be considered algorithmically random the proportions of 0 and 1 must be very nearly equal. One can say

Algorithms, Information, and Chance

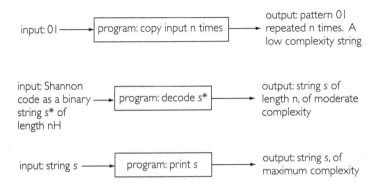

Figure 4.2 Representations of computer algorithms for strings of low, moderate, and maximal algorithmic complexity.

more. Every possible block of 0 and 1 of length k in a random sequence must also appear with roughly equal frequency for each $k = 1, 2, \ldots$; otherwise patterns can be detected that can be used to shorten the description, in a similar manner to a Shannon coding. For infinite strings you then reach a familiar conclusion: Random sequences must define *normal numbers*. Algorithmic randomness is therefore consonant with randomness in the usual sense of maximum entropy and of normal numbers.

Incidently, though the sine qua non of randomness of a string is that it be the binary representation of a *normal* number, not every normal number is random. In Chapter 1 you encountered the only well-documented example of such a number,

What Is Random?

due to D. Champernowne in 1933, and it is characterized by a simple description of its binary expansion: Choose each digit singly, then in pairs, then in triplets, and so on. At the nth step, list in order all possible 2^n strings of length n, giving rise to the sequence 0100011011000001.... Any sufficiently long but finite segment of this number cannot be random, since the algorithm for generating it is fairly short.

Remarkably, if every book written were to be coded in binary form, each of them would sooner or later surface within a long enough segment of Champernowne's number, since every possible string eventually appears and reappears indefinitely. Your favorite novel or recipe or even the book you are currently reading will pop up somewhere along the way. Champernowne's number contains the vast storehouse of written knowledge in a capsule.

Randomness in the information-theoretic sense of maximum entropy was framed in Chapter 2 in terms of the probabilities of the source symbols. By contrast, the complexity definition of randomness makes no recourse to probability, and it depends only on the availability of a mechanical procedure for computing a binary string. Moreover, the complexity definition is independent of the provenance of a string s. It looks only at the string itself, as merely a succession of digits. In this way it overcomes the embarrassment of having to contemplate a not so very random looking output of some allegedly random mechanism; remember that a Bernoulli ½-process can crank out a string of all zeros or a totally unpatterned string with the same insouciance.

Algorithms, Information, and Chance

In the previous chapter it was mentioned that a random string is random when viewed in reverse. This can be seen from the complexity definition as well: If s taken backwards can be reproduced by a shorter string $s\star$, then the relatively short program "read $s\star$ in reverse" will reconstruct s, and this contradicts the putative randomness of s.

A similar argument shows that the minimal program $s\star$ must itself be random. Otherwise, there is a shorter string $s\star\star$ that reproduces $s\star$, and a new program can be created that concatenates $s\star$ and $s\star\star$ with the command "use $s\star\star$ to generate $s\star$ and follow the instructions of $s\star$". After the machine executes $s\star\star$, a few more bits of program are needed to instruct the computer to position itself at the beginning of $s\star$ and continue until s appears. The new program that spawns s is shorter than $s\star$, if $s\star$ is sufficiently long to begin with, and so $s\star$ is not minimal. This contradiction establishes that $s\star$ must indeed have have maximum complexity (see Figure 4.3).

Randomness of s has been defined by the algorithmic complexity of s matching the length of s. This can be made more precise by stipulating that a string of length n is *c-random* if its algorithmic complexity is greater than or equal to $n - c$ for some small integer c. There are at most 2^i distinct programs of length i, for $i = 1, 2 \ldots, n - c - 1$, and therefore at most $2 + 2^2 + \ldots + 2^{n-c-1} = 2^{n-c} - 2$, or fewer than 2^{n-c}, programs that describe strings of length n having complexity less than $n - c$ (Appendix A explains how this sum is arrived at). Since there are 2^n strings of length n, the fraction of strings that are not c-random is less than $2^{n-c}/2^n = 1/2^c$. With $c = 10$, for instance,

What Is Random?

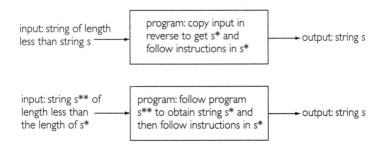

Figure 4.3 Representations of computer programs for establishing that the converse of a random string is again random, and to show that a minimal program $s^\star$ is itself random.

fewer than one in a thousand strings are not c-random, since $1/2^{10} = 1/1024$. From now on the randomness of long strings is identified with c-randomness in which c is negligible relative to the length of the string. With this proviso, *most strings are random because their algorithmic complexity is close to their length.*

There is a link between algorithmic complexity and the *Janus algorithm* of the previous chapter. Assume, for simplicity, that the Janus iteration begins with the seed value u_0 equal to zero, namely a string of zeros, and let us watch as it successively computes n digits of Bernoulli $1/2$-trials from right to left. Initially, the unknown string is one of 2^n equally probable strings of length n, and its information content is n bits. But after k iteration the information content of the remaining portion of the string is $n - k$, since only $n - k$ bits remain to be

Algorithms, Information, and Chance

specified. However, if the string is incompressible, the first k bits represent a block of complexity roughly k, and as k increases, for $k = 1, 2, \ldots n,$ the uncertainty diminishes while the complexity grows. Now apply the inverse to Janus to erase the digits created, one step at a time, from left to right. After k erasures the information content rises to k, since k bits have been lost, but the complexity of the remaining portion of the string has dwindled to roughly $n - k$. The sum of the two terms, entropy and complexity, remains constant at each step.

There is a duality here between the increase in entropy in one direction and an increase in complexity in the opposite direction; *information content and complexity are seen as complementary ways of representing randomness* in an individual string.

If Janus is modified to produce digits from Bernoulli p-trials then some redundancies can possibly be squeezed out, and so after k steps the remaining uncertainty is only $(n - k)H$, where H is the entropy of the source. The complexity of the remaining portion is roughly kH using a Shannon encoding, as you saw above. Once again, the sum of the two terms is nearly constant at each iteration.

The duality between ignorance now and disorder later is an idea that finds resonance in the recent work of physicist W. H. Zurek in connection with the Second Law of Thermodynamics. Zurek points out that, like the engine powered by Szilard's demon that we met in the previous chapter, measurements decrease entropy by reducing the uncertainty of the future but increase algorithmic complexity by leaving behind a disordered trail of digits. Between one measurement and the

What Is Random?

next, the average change in entropy in one direction equals the average change in complexity in the opposite direction, and so these quantities are conserved at each step.

The self-reflection of Janus mirrors the hallucinatory nature of randomness in which ignorance begets disorder, which begets ignorance. The Argentine writer Jorge Luis Borges captured this narcissism in an essay in which he taunts us with "Why does it disturb us that the map is included in the map and a thousand and one nights in the book of the *Thousand and One Nights?* Why does it disturb us that Don Quixote be a reader of the *Quixote* and Hamlet a spectator of *Hamlet?* I believe I have found the reason: these inversions suggest that if the characters of a fictional work can be readers or spectators, we, its readers or spectators, can be fictitious." A more lighthearted version of these musings is captured by the *New Yorker* cartoon in which an announcer steps through the parted curtain and notifies the theatergoers that "tonight the part normally played by the audience will be played by the actors playing the part of the audience."

Figure 4.4, a variant of Figure 3.10, gives a schematic if somewhat prosaic representation of Janus's self-reflection.

With the complexity definition in hand randomness of a binary string can now be understood in three senses:

- A string is random if each digit is generated by some mechanism in an unpredictable manner. The randomness resides in the disorder of the generating process. The only way the string can be reproduced is for a monkey

Algorithms, Information, and Chance

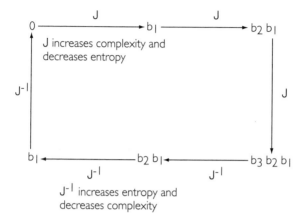

Figure 4.4 Schematic representation of how entropy and complexity as two sides of the same Janus-faced coin. This figure is a variant of Figure 3.10 and uses the same notation adopted there.

working at a binary switch accidently to hit upon its pattern.
- A string is random because it is completely unanticipated; its entropy is maximal. Recognition of the string elicts total surprise.
- A string is random because no prescribed program of shorter length can regurgitate its successive digits. Since it is maximally complex in an algorithmic sense, the string cannot be reproduced except by quoting itself, a futile circularity.

What Is Random?

Algorithmic Complexity and Undecidability

The notion of algorithmic complexity was defined in terms of a program executed on a computer. To make further progress with this idea and its connection to randomness it is necessary to clarify what is meant by a computer.

In 1936, well before the advent of electronic computers in the 1940s, the mathematician Alan Turing reduced the idea of computation to its bare bones by envisioning a simple device that carries out in rote fashion a finite number of commands, one at a time, following the precise and unambiguous steps of an algorithm, or in computer jargon, a *program*. Turing's conceptual device incorporated the essential ingredients of a modern computer and was a precursor to the actual computing machines that were to be built a decade later.

His imaginary machine consists of a scanner that moves along a tape partitioned into cells lined up in a single row along a strip. Each cell contains either a one or a zero, and a portion of the tape consists of the binary expansion of some number that represents the input data, one digit per cell. Depending on what the scanner reads on the first cell it is positioned at, it carries out one of a small set of moves. It can replace what it reads with either a 0 or 1, or it can move one cell right or left, or it can go to one of the finite number of steps in the algorithm and carry out whatever instruction it finds there, and this might include the command "stop." The finite set of instructions is the machine's program.

Algorithms, Information, and Chance

A simple example of a Turing computation is the following three steps:

- step 1 Move one cell to the right.
- step 2 If the cell contains a 0, go to step 1.
- step 3 If the cell contains a 1, go to step 4.
- step 4 Stop.

Here is a key observation: The instructions in the program may themselves be coded in binary form. There are seven code words:

code	instruction
000	write 0
001	write 1
010	go left
011	go right
1010...01	go to step i if 0 is scanned
	(there are i zeros to the right of the leftmost 101)
1011...10	go to step i if 1 is scanned
	(there are i ones to the right of the leftmost 101)
100	stop

What Is Random?

An example of the fifth instruction is 10100001, which says "go to step 4 if 0 is scanned." The simple program of four steps given earlier can therefore be coded as the string 0111010110111110100, since 011 is step 1, 10101 is step 2, 10111110 is step 3 and 100 is the final step "stop." In effect, if a 1 is encountered after step 1 then the program moves to step 4, and then stops.

The translation of a program into binary form is evidently a prefix code and is therefore immediately decipherable by reading left to right. Suppose that a Turing machine, labeled T, carries out some computation on a given input data string and stops when some output string s appears on the tape. Turing conceived another machine, labeled U, that systematically decodes the binary instructions in the program for T and then mimics the action of T on its input data. This mechanism, the *universal Turing machine,* operates on a tape whose initial configuration is the coded program for T followed by the input data. All that U has to do is unwaveringly decode each command for T and then scuttle over to the input string and carry out the operation that T would have itself performed. It then returns to the next command in T's program and repeats this scurrying back and forth between program and data. T's program is stored on the tape as if it were input data for U, and it passively cedes its role to U.

The computers in commercial use conform to the same general scheme: You write a program in some language like Fortran, and the computer then interprets these commands using its own internal language, that forges an output using the

Algorithms, Information, and Chance

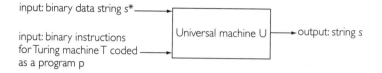

Figure 4.5 Schematic of a universal Turing machine U that acts on the binary instructions of a Turing machine T coded as a binary program p. Turing machine U simulates the behavior of T on an input data string $s\star$ to produce an output string s. This may be expressed symbolically as $U(p, s\star) = T(s\star)$.

binary logic inscribed on its chips. Figure 4.5 is a schematic representation of a universal Turing machine.

One of the benefits of a universal Turing machine U is that it removes an ambiguity that lurked in the background of the definition of algorithmic complexity of a string s. The length of the shortest program $s\star$ that replicates s on a given computer depends on the idiosyncrasies of that machine's hardware. To remove any dependency on the particular machine in use, it is convenient to define the complexity in terms of a universal computer that simulates any other device. This, of course, introduces some additional overhead in terms of the instructions needed to code U's internal interpreter. This overhead increases the program by a fixed number of bits but is otherwise independent of the string s that is the desired output. Any two machines that compute s have programs whose

What Is Random?

lengths can differ by at most a constant number of bits, since each of them can be simulated by the same U, and so the complexity of s is effectively machine-independent. The additional "overhead" becomes less significant as s gets longer, and it is useful to think of lengthy strings when talking about complexity so that machine dependency can be neglected.

Since Turing's original work it has become an entrenched dogma, the *Church–Turing thesis* (so called because Turing's work shadowed that of the American logician Alonzo Church that was carried out independently and at about the same time), stating that anything at all computable using some algorithm, either mechanically or by a human, can also be performed by a universal Turing machine. This stripped-down, almost toy-like, caricature of a computer has the remarkable property of being able to realize, in principle at least, any program whatever. We are talking here of the ability to carry out the steps of any program, but an actual Turing computation, if it could be realized in practice, would be woefully slow.

Not every Turing computation halts when presented by some input string. The fourstep program given above halts after a finite number of moves if the input contains some cell with a one on it; for an input consisting solely of zeros it grinds away without ever stopping. A glance at this program reveals immediately whether it stops or not for a given input, but in general it is not obvious when or whether a program will halt. Turing established the startling result that there is no program P crafty enough to determine whether any given program T will eventually halt when started with some input data string. The in-

Algorithms, Information, and Chance

ability to decide whether a Turing computation will halt is closely linked to Gödel's notorious incompleteness theorem, and it is mirrored by an equally striking assertion regarding algorithmic complexity due to Gregory Chaitin. I relegate Turing's argument to the technical notes[4.1] and move instead directly to what Chaitin showed.

Let us regard a proof of an assertion as a purely mechanical procedure using precise rules of inference starting with a few unassailable axioms. This means that an algorithm can be devised for testing the validity of an alleged proof simply by checking the successive steps of the argument; the rules of inference constitute an algorithm for generating all the statements that can be deduced in a finite number of steps from the axioms. There is, in effect, a Turing program that can work through the chain of inferences to test the validity of an alleged proof. If the axioms and proofs are coded as binary strings, the program begins by checking binary proof strings of length 1, then all such strings of length 2, then those of length 3, and so on, and whenever a valid proof is encountered, namely a proof that follows from the axioms using the given rules of inference, it prints it out. Eventually, any valid proof will be found in this manner by the Turing program.

Consider now the claim that there is a proof of the statement "for any positive integer n there is a binary string whose complexity exceeds n." The Turing program is an algorithm that tests all possible proofs in order of increasing length until it finds the first one able to establish that some string has complexity greater than n. It then writes out this string and stops.

What Is Random?

The program has a length of k bits, where k is the fixed length of the binary string that encodes the particular choice of axioms as well as the instructions required to check the details of the proof, plus about $\log n$ bits that are needed to code the integer n in binary form (see Appendix C to verify the last assertion). Though $\log n$ increases as n gets larger, it does so more slowly than n (see technical note 4.3), and therefore, for n large enough, n will eventually exceed the length of the program, namely $k + \log n$. Here you have a contradiction: The program supposedly *computes the first string that can be proven to have complexity greater than n, but this string cannot possibly be calculated by a program whose length is less than n*. The contradiction can only be avoided if the program never halts!

Chaitin's proof is related to a paradox posed by Oxford librarian G. Berry early in the twentieth century that asks for "the smallest positive integer that cannot be defined by an English sentence with fewer than 1000 characters." Evidently, the shortest definition of this number must have at least 1000 characters. However, the sentence within quotation marks, which is itself a definition of the alleged number, is less than 1000 characters in length! This paradox appears to be similar to that of "the first string that can be proven to be of complexity greater than n." Although the Berry paradox is a verbal conundrum that seems not to be resolvable, in Chaitin's case the paradox dissolves by simply observing that no such proof exists. That is, it is not possible to ascertain whether a given binary string has complexity greater than some sufficiently large n, for if such a proof existed, then as we saw, there would be an algorithm that uses fewer than n bits to carry out the proof, and that contradicts the meaning of complexity.

A paraphrase of Chaitin's result is that there can be *no formal proof that a sufficiently long string is random* because there is no assurance that its complexity can ever be established, since the complexity of a random string is comparable to its length. No matter how one devises the axioms and an algorithm for testing valid proofs, there is always some long enough string that eludes a proof of randomness, even though it may, in fact, be random.

In response to the persistent question "*Is it random?*" the answer must now be "probably, but I'm not sure"; "probably" because most numbers are, in fact, random, as we have seen, and "not sure" because the randomness of long strings is essentially undecidable. A monkey is likely to generate a long string that is in fact random, but it is unlikely that we would be able to recognize it as such.

Algorithmic Probability

Suppose that a universal Turing machine U operates on a random input. More specifically, U is fed a string obtained from Bernoulli ½-trials, and it then tries to interpret the randomly generated digits as if they constituted the binary instructions of some program. Let the symbol Ω denote the probability that U halts when it is fed a random input; we call Ω the *halting probability*. It is the sum, over all programs that halt, of the individual

What Is Random?

probabilities of actually picking those very programs. If U halts after l steps for a program string p, then all the potentially infinite strings of Bernoulli ½-trials that begin with p must also halt, because if s denotes the remaining sequence of digits, the concatenation of p and s necessarily halts. The probability of halting on all such strings that begin with p is therefore 2^l (the optional details of how this probability is arrived at are found in technical note 4.2). Since p evidently cannot be the prefix of any other halting program p', it follows that U halting with p' is an event that is disjoint from U halting with p. (See technical note 4.2.) Of course, there may be several halting programs of the same length, and each of these makes a separate contribution to the halting probability. The probabilities of the disjoint events "U halts for a given program" are summed to arrive at the single probability Ω that U halts at all. I leave to the technical notes any assurance that the sum is a number between zero and one (technical note 4.2), as befits a probability, and concentrate instead on discussing some of the startling properties of Ω.

The halting probability Ω has a binary expansion $a_1 a_2 a_3 \ldots$, and the first n bits define a number Ω_n. The remaining digits of Ω sum to something less than or equal to $\frac{1}{2}^n$ (see Appendix A for an explanation of why this is so). We have already encountered a nonhalting program, and so U does not halt for every program; this shows that Ω is actually less than one. Therefore, Ω is a number greater than or equal to Ω_n and less than $\Omega_n + \frac{1}{2}^n$.

From Ω_n it is possible to decide the halting or nonhalting of all programs of length not exceeding n. List all possible ran-

Algorithms, Information, and Chance

dom program strings as {0, 1, 00, 01, 10, 11, 000, 001, 010, ...} and then begin a systematic search of each in turn, starting with the smallest, to see whether U halts or not on the given string. However, an impasse is reached when you come to the first program that doesn't halt, since this effectively blocks all subsequent programs from being checked. To avoid getting stuck this way, a *dovetailing procedure* is employed instead, in which U carries out the first step of the first program followed by one step of each of the first two programs, then one step of each of the first three programs, then one step of each of the first four programs, and so on. In this manner U eventually runs all possible programs, and if any of these halt, they contribute to the probability Ω_n by an amount determined by their length as discussed above, while those that do not halt contribute nothing. The dovetailing ensures that any halting program will appear sooner or later as a particular input string to U.

When enough programs have halted, the sum of their respective probabilities will equal or exceed Ω_n for the first time. Some of these programs are longer than n bits and some shorter, and although there may still be other programs that will halt later, there cannot be any additional programs of length l less than or equal to n that can halt! The reason is that such a program would contribute an additional probability $\frac{1}{2}^l$, which is at least $\frac{1}{2}^n$, giving a total probability that equals or exceeds $\Omega_n + \frac{1}{2}^n$, and this not admissible (see Figure 4.6 for a pictorial representation of this argument). Therefore, all programs of length at most n that have not yet halted will never

What Is Random?

halt, and the n bits of Ω_n encapsulate the information needed to decide the halting or nonhalting of all programs of length at most n. The dovetailing procedure may be quite lengthy, but it is also of finite duration, since in due course the probabilities of the halted programs must sum to a number greater than Ω_n, and when this happens, the scales tip and the procedure ends.

Each program that halts will have been noted, and this includes all those of length not exceeding n. This is quite extraordinary, since many unresolved problems in mathematics can be coded as programs whose length is less than a sufficiently large n. For example, the unproven conjecture that every even number is the sum of two prime numbers (a number is prime if its only divisors are 1 and itself) could be decided in principle with a program that searches each even number in turn, checking to see whether there are two primes less than n whose sum is n. If there is no such number, then the program will never halt. Therefore a knowledge of whether the program halts decides the truth or falsity of the conjecture.

The drawback in this entire discussion is that we don't know Ω_n, and indeed the oracle Ω_n is random. To see this, recall that the complexity of a random string s, namely the length of the shortest program $s^\star$ that reproduces s, is comparable to the length of s. Among the halting programs are those that output random strings of length up to and including n. The algorithmic complexity of Ω_n *cannot be less than n*, for otherwise there would be a program of length less than n that generates Ω_n and, so doing, would (in the manner of Figure 4.3b) also generate all programs that produce random strings of length n;

Algorithms, Information, and Chance

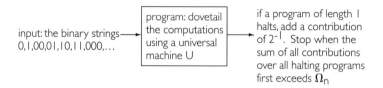

Figure 4.6 Illustration of how the first n bits of the binary string Ω determine all halting programs of length not exceeding n. A universal Turing machine U acts on random program strings of increasing length. Each halting program p of length l contributes a halting probability 2^{-l}. The computation stops when a sufficient number of programs have halted so that the sum of the individual halting probabilities first exceeds Ω_n.

but this is impossible by the definition of algorithmic complexity. It follows that Ω_n must, in fact, be random. As Charles Bennett wistfully remarks, "Ironically, although Ω cannot be computed, it might be generated accidently by a random process, such as a series of coin tosses or an avalanche that left its digits spelled out in the pattern of boulders on a mountainside. The first few digits of Ω are probably already recorded somewhere in the universe."

The number Ω_n reminds me of the cult movie "π", in which there is a frantic search for a mystical number of fixed length whose digits would reveal some long-sought secret. From the divination of portents among the ancients, to the cabalistic interpretation of sacred texts in the Middle Ages, to the

What Is Random?

succinct equations of quantum theory in our own time, there is a long tradition of trying to fathom the unknown by finding an omniscient string of numbers or symbols. Ω_n is such a number, since it achieves the goal of answering a large number of opaque mathematical questions with a not very long string of digits. But Ω_n is random, and so, though we know it exists, it remains inscrutable. Its wisdom, like Borge's Library of Babel, is useless.

The succinctness of Ω may be appreciated when compared to the number K, whose binary representation is defined in terms of the enumeration of programs {0, 1, 00, 01, 10, 11, 000, ... } by setting the ith digit to 1 or 0 depending on whether the ith program in the list does or does not halt. The number of programs of length less than n does not exceed 2^n (Appendix A), and so the first 2^n digits of K convey the same information as the first $n-1$ digits of Ω. As Bennett explains, "Many instances of the halting problem are easily solvable or reducible to other instances," and this redundancy ensures that K is not random.

The halting probability Ω was defined in terms of all random program inputs that cause a universal Turing machine to halt. A slight modification of Ω leads to the probability P_s that a random program will output a particular string s; choose P_s to be the sum, over all programs that output s, of 2^{-l}, where l is the length of the program that yields s. As before, the term 2^{-l} is the probability that the program p of length l is chosen among the 2^l equally probable strings having that length. It may be shown, although I do not do so here, that P_s is effectively the same as

Algorithms, Information, and Chance

$2^{-C(s)}$ where $C(s)$ denotes the algorithmic complexity of s. This means that low-complexity strings are more likely than those of high complexity, and it ensures that if one is given two strings s_1 and s_2, then the *shorter one is the more probable of the two*.

It is part of the lore of science that the most parsimonious explanation of observed facts is to be preferred over convoluted and long-winded theories. Ptolemaic epicycles gave way to the Copernican system largely on this premise, and in general, scientific inquiry is governed by the oft-quoted dictum of the medieval cleric William of Occam that "nunquam ponenda est pluralitas sine necesitate," which may be paraphrased as "choose the simplest explanation for the observed facts." This fits in with the conclusion reached in the previous paragraph that if two hypotheses are framed as binary strings, the most likely hypothesis is the shorter one.

The mathematician Laplace has already been quoted in the first chapter regarding the need to distinguish between random strings and "those in which we observe a rule that is easy to grasp," a prescient remark that anticipates algorithmic complexity. Laplace preceeds this statement by asserting that

> If we seek a cause whenever we perceive symmetry, it is not that we regard a symmetrical event as less probable than the others but, since this event ought to be the effect of a regular cause or that of chance, the first of these suppositions is more probable than the second. On a table we see letters arranged in this order, *constantinople,* and we judge that this arrangement is not the result of chance, not because it is less possible than the others, for if this word were

What Is Random?

> not employed in any language we would not suspect that it came from any particular cause, but with the word in use among us, it is incomparably more probable that some person has arranged the aforesaid letters than that this arrangement is due to chance.

Later he continues, "Being able to deceive or to have been deceived, these two causes are as much more probable as the reality of the event is less."

A string s of length n generated by a *chance* mechanism has a probability $½^n$ of occurring, but a *cause* for s would be a much shorter string $s^\star$ of length m that outputs s on a universal computer U. The probability of getting $s^\star$ is $½^m$ out of all equally likely programs of length m, where m is, of course, the complexity $C(s)$. The ratio of 2^{-m} to 2^{-n} is 2^{n-m}, and whenever m, namely $C(s)$, is much less than n, it is then 2^{n-m} times *more probable that s arose from a cause than from chance*. Short patterns are the most convincing, perhaps because their brevity gives them the feel of truth. That may be why we are so attracted to the pithy assesrtion, the succinct poem, the tightly knit argument. The physicist Werner Heisenberg wrote, "If nature leads us to mathematical forms of great simplicity and beauty . . . we cannot but help thinking that they are true."

[5]

The Edge of Randomness

And so one may say that the same source of fortuitous perturbations, of "noise," which in a nonliving (i.e., nonreplicative) system would lead little by little to the disintegration of all structure, is the progenitor of evolution in the biosphere and accounts for its unrestricted liberty of creation, thanks to the replicative structure of DNA: that registry of chance, that tone deaf conservatory where the noise is preserved along with the music.

Jacques Monod, *Chance and Necessity*

What Is Random?

VALENTINE: It's all very, very noisy out there. Very hard to spot the tune. Like a piano in the next room, it's playing your song, but unfortunately it's out of whack, some of the strings are missing, and the pianist is tone deaf and drunk— I mean, the noise! Impossible!

HANNAH: What do you do?

VALENTINE: You start guessing what the tune might be. You try to pick it out of the noise. You try this, you try that, you start to get something—it's half baked, but you start putting in notes that are missing or not quite the right notes . . . and bit by bit . . . the lost algorithm!

Tom Stoppard, *Arcadia*

Between Order and Disorder

Up to now I have been trying to capture the elusive notion of chance by looking at binary strings, the most ingenuous image of a succession of sensory events. Since most binary strings cannot be compressed, one would conclude that randomness is pervasive. However, the data streams of our consciousness do, in fact, exhibit some level of coherence. The brain processes sensory images and unravels the masses of data it receives, somehow anchoring our impressions by allowing

The Edge of Randomness

patterns to emerge from the noise. If what we observe is not entirely random, it doesn't mean, however, that it is deterministic. It is only that the correlations that appear in space and time lead to recognizable patterns that allow, as the poet Robert Frost puts it, "a temporary stay against confusion." In the world that we observe there is evidently a tension between order and disorder, between surprise and inevitability.

I wish to amplify these thoughts by returning to binary strings, thinking of them now as encoding the fluctuations of some natural process. In this context the notion of algorithmic complexity that preoccupied us in the previous chapter fails to capture the idea of complexity in nature or in everyday affairs as it is usually perceived. Consider whether the random output of a monkey at a keyboard is more complex than a Shakespearean sonnet of the same length. The tight organizational structure of the sonnet tells us that its algorithmic complexity is less than the narration produced by the simian. Evidently, we need something less naive as a measure of complexity than simply the length of the program that reproduces the sonnet. Charles Bennett has proposed instead to measure the complexity of a string as the *time required, namely the number of computational steps needed,* to replicate its sequence from some other program string. Recall the algorithmic probability P_s that a randomly generated program will output s. It was defined in Chapter 4 as a sum over all programs that generate s. Evidently, P_s is weighted in favor of short programs, since they contribute more to the sum than longer ones. If most of P_s is contributed by short programs for which the number of steps

What Is Random?

required to compute s is large, then the string s is said to have *logical depth*.

The lengthy creative process that led to Shakespeare's sonnet endows it with a logical depth that exceeds by far the mindless key tapping of the monkey, even though the sonnet has less algorithmic complexity because of redundancies, some obvious and others more subtle, that crop up in the use of language. Although the information content of the sonnet may be limited, it unpacks a one-of-a-kind message that is deep and unexpected.

Allowing that a long random string s can be replicated only by a program of roughly the same length as s, the instructions "copy s" can nevertheless be carried out efficiently in a few steps. Similarly, a totally ordered string such as 01010101 ... requires only a few computational steps to execute "copy 01 n times," since a simple copy instruction is repeated many times over and again. Clearly, strings like these that entail very large or very small algorithmic complexity are shallow in the sense of having meager logical depth: Little ingenuity is needed to execute the steps of the program. Therefore, strings that possess logical depth must reside somewhere between these extremes, *between order and disorder*. As Charles Bennett put it, "The value of a message thus appears to reside not in its information (its absolutely unpredictable parts), not in its obvious redundancy (verbatim repetitions, unequal digit frequencies), but rather in what might be called its buried redundancy—parts predictable only with difficulty." The number K of the previous chapter provides an illuminating example of logical depth. You may recall that K had its binary digits defined in terms of an enumeration {0, 1, 00, 01, 10, 11, 000 . . . } of all programs, with the ith digit equal to one or zero according to whether the ith program in the list does or does not halt. It was seen that the first 2^n digits of K are quite redundant and contain the same information as roughly the first n digits of Chaitin's random number Ω, establishing, in Bennett's words, that "Ω is a shallow representation of the deep object K."

Another striking example of logical depth is provided by DNA sequences. This familiar double-stranded helix is made up of nucleotides consisting of sugars and phosphates hooked to one of four different bases designated simply as A, C, G, T. Each triplet from this alphabet of symbols codes for one of the twenty amino acids that are linked together to form proteins. Some of these proteins are enzymes that, in a self-referential way, regulate the manner in which DNA unfolds to replicate itself, and other enzymes regulate how DNA transcribes its message to make more proteins.

Writing each of the four symbols in binary form as 00, 01, 10, 11 exhibits DNA as a long binary string (about 3 billion symbols in humans). Since there are $4^3 = 64$ possible triplets of nucleotides A, C, G, T, called codons, and only 20 amino acids, there is some duplication in the way transcription takes place. Moreover, some fragments of the DNA strand repeat many times, and there also appear to be long-term correlations between different portions of the string, and this suggests that there is considerable redundancy. On the other hand, some codon sequences appear to be junk, having no recognizable

What Is Random?

role, possibly caused by chance accretions over the span of evolutionary time. It follows that DNA lies betwixt randomness and structure, and its logical depth must be substantial, since the evolution of the code took place over several million years. A DNA string s can therefore be replicated by a shorter program string $s^\star$, but the blueprint of the more succinct code $s^\star$ is likely to be complicated, requiring a lot of work to unpack it into a full description of the string s.

The genetic machinery of a cell provides not only for the copying of DNA and, indirectly, for its own replication, but it controls the onset of an organism's growth through the proteins coded for by DNA. The cells combine to form organisms, and the organisms then interact to form ecosystems. The mechanical process of DNA replication is disrupted from time to time by random mutations, and those mutations function as the raw material for natural selection; evolution feeds on the fortuitous occurrence of mutations that favor or inhibit the development of certain individuals.

Chance also intrudes beyond the level of genes as organisms have unexpected and sometimes disruptive encounters with the world around them, affecting the reproduction and survival of individuals and species. Some of these contingent events, what physicist Murray Gell-Mann calls "frozen accidents," have long-term consequences because they lock in certain regularities that persist to provide succeeding organisms, individually and collectively, with characteristics that possess some recognizable common ancestry. The accumulation of such frozen accidents gives rise to the complexity of forms that we observe around us. The helical structure of DNA and of organic forms like snails may be a consequence of such accidents. Jacques Monod expressed the same idea most evocatively as "randomness caught on the wing, preserved, reproduced by the machinery of invariance and thus converted into order, rule, necessity."

Some accidents of nature permit existing parts to be adapted to new functions. Paleontologist Stephen Jay Gould comments on how happenstance provides an opportunity for selection when he argues that the complexity of forms is due to "poor fit, quirky design, and above all else, redundancy.... Pervasive redundancy makes evolution possible. If animals were ideally honed, with each part doing one thing perfectly, then evolution would not occur, for nothing could change and life would end quickly as environments altered and organisms did not respond." Fully formed parts are unlikely to be made from scratch but are a product of nature's fortuity. Catastrophic events, such as large meteor impacts, can also alter the course of evolution by extinguishing some species and bestowing a selective advantage to other, previously marginal, dwellers.

Redundancy in DNA and in the organisms it spawns makes them less fragile to disruptions. The same is true at the level of ecosystems, which apparently can achieve species diversity because of spatial heterogeneity and fluctuations in the environment. Disturbance of an ecological community allows new species to colonize areas that would normally be inhabited by a more agressive competitor, and sporadic incursions

What Is Random?

such as floods, storms, or fires can leave in their wake a patchy landscape coinhabited by more species than would otherwise be possible in a more stable environment in which predation and competition would ensure the dominance of just a few species. The ecologist G. Evelyn Hutchinson once pondered "the paradox of the plankton" whereby a number of competing species of plankton are able to coexist, rather than one species surviving at the expense of the others, and he similarly concluded that this was possible because turbulence in the waters dislodges the community structure from equilibrium.

Whether it be at the level of cells, organisms, or ecosystems, chance and order comingle to unfold the vast panorama of the living world. This underscores the utility of randomness in maintaining variability and innovation while preserving coherence and structure.

A number of thinkers have gone further in suggesting how some semblance of order and coherence can arise from irregular and accidental interactions within biological systems. The idea is that large ensembles of molecules and cells tend to organize themselves into more complex structures, hovering within the extremes of, on one hand, totally random encounters where turmoil reigns and no order is possible and, on the other, tightly knit and rigidly regulated interactions where change is precluded. Cells and organisms in isolation, shut off from the possibility of innovation, veer toward decay and death. Biologist Stuart Kauffman believes that "selection . . . is not the sole source of order in biology, and organisms are not just tinkered-together contraptions, but expressions of deeper natural laws ... Profound order is being discovered in large, complex, and apparently random systems. I believe that this emergent order underlies not only the origins of life itself, but much of the order seen in organisms today."

For physicist Per Bak, emergent order occurs not only in the realm of biological phenomena, but it is rampant in the worlds of physical and social experience: "Complex behavior in nature reflects the tendency to evolve into a poised critical state, way out of balance, where minor disturbances may lead to events, called avalanches, of all sizes.... The evolution to this very delicate state occurs without design from any outside agent. The state is established solely because of the dynamical interactions among individual elements of the system: the critical state is *self-organized.*"

Kauffman underscores these sentiments when he says that

> Speciation and extinction seem very likely to reflect the spontaneous dynamics of a community of species. The very struggle to survive, to adapt to small and large changes ... may ultimately drive some species to extinction while creating novel niches for others. Life, then, unrolls in an unending procession of change, with small and large bursts of speciations, small and large bursts of extinctions, ringing out the old, ringing in the new ... these patterns ... are somehow self organized, somehow collective emergent phenomena, somehow natural expressions of the laws of complexity.

What Is Random?

Chemist Ilya Prigogine had, even earlier, given vent to similar ideas when he advocated a view of nature far from equilibrium in which evolution and increasing complexity are associated with the self-organization of systems that feed on the flux of matter and energy coming from the outside environment "corresponding to a delicate interplay between chance and necessity, between fluctuations and deterministic laws." He contrasts these pools of decreasing entropy with the irrevocable degradation and death implied by the Second Law of Thermodynamics. Complex structures offset the inevitable downhill slide into disintegration by dumping their own decay into the open system that nourishes them. Recall the Szilard demon of Chapter 3 who is able to decrease entropy and increase order, bit by bit, by accumulating a record of junk digits that ordinarily need to be erased to restore the loss in entropy. But if these digits are unceremoniously wasted into the environment, a pocket of order is created locally even as disorder is increased globally.

The thesis of self-organized complexity is a controversial idea that finds its most ardent voice today at the Santa Fe Institute, in New Mexico, and it is not my intention to engage in the ongoing polemic regarding the validity of this and competing ideas of complexity that are vexing these thinkers. This is best reviewed in a spate of books that have appeared in the last few years, not the least of which is Bak's effort in *How Nature Works*. Instead, I accept this idea as a provocative metaphor of how chance and order conspire to provide a view of complexity in nature, and in the artifacts of man.

The Edge of Randomness

That randomness gives rise to innovation and diversity in nature is echoed by the notion that chance is also the source of invention in the arts and everyday affairs in which naturally occurring processes are balanced between tight organization, where redundancy is paramount, and volatility, in which little order is possible. One can argue that there is a difference in kind between the unconscious, and sometimes conscious, choices made by a writer or artist in creating a string of words or musical notes and the accidental succession of events taking place in the natural world. However, it is the perception of ambiguity in a string that matters, and not the process that generated it, whether it be man-made or from nature at large.

In Chapter 2 we saw that the English language is neither completely ordered, which would render it predictable and boring, nor so unstructured that it becomes incomprehensible. It is the fruitful interplay between these extremes that gives any language its richness of nuance. The same is true of the music of Bach or Mozart, to mention just two composers, which is poised between surprise and inevitability, between order and randomness. Many architects attempt to combine wit with seriousness of design to create edifices that are playful and engaging while meeting the functional requirements dictated by the intended use of these buildings.

In a book about mystery and romance John Cawelti states, "if we seek order and security, the result is likely to be boredom and sameness. But, rejecting order for the sake of change and novelty brings danger and uncertainty.... the history of culture can be interpreted as a dynamic tension between these two basic impulses . . . between the quest for order and the flight from ennui."

A final example of how chance intrudes to provide an opportunity for novelty and complexity and the formation of patterns far from equilibrium is based on the cliché, popularized by John Guare in his play *Six Degrees of Separation,* that everyone is connected to everyone else in the world through at most six intermediate acquaintances. A study by mathematicians Duncan Watts and Steven Strogatz shows that an ensemble of entities tightly woven into parochial clusters can rapidly expand into a global network as soon as a few links are randomly reconnected throughout the network. Examples abound to show that structures as diverse as the global community of humans, electric power grids, and neural networks have all evolved to reside somewhere between a crystalline structure of local connectedness and random disarray.

All these examples are reminiscent of a delightful watercolor by the eighteenth-century artist Pietro Fabris, one of a series used to illustate a work called *Campi Phlegraei* by the urbane scholar and diplomat Sir William Hamilton, British envoy to the then Kingdom of Naples. In the watercolor Hamilton is leaning on his staff below the crater of Vesuvius, viewing the sulfurous pumice being hurled haphazardly from the belching vent, reflecting, it seems, on the delicate balance between the tumult and anarchy of the untamed volcano and the unruffled azure sky beyond, between the capriciousness of ordinary life and the world of reason, an exquisite portrayal of life poised between order and disorder.

The Edge of Randomness

Self-Similarity and Complexity

"Clouds are not spheres, mountains are not cones, coastlines are not circles, and bark is not smooth, nor does lightning travel in a straight line," asserts Benoit Mandelbrot in his influential and idiosyncratic book on fractal geometry. A fractal structure is not smooth or homogeneous. When one looks at a fractal more and more closely, greater levels of detail are revealed, and at all the different scales of magnification, the fractal structure exhibits more or less similar features. An example often mentioned is the coastline of Britain, whose perimeter increases the more closely it is examined. On a scale of 10 kilometers it has a perimeter that misses many tiny inlets and coves, but these make their appearance when measured with a 10-meter measuring stick, and all the additional indentations uncovered at this scale serve to increase the length. The crinkliness of the shoreline continues until one gets down to the dimension of individual pebbles. Another illustration is familiar from Chapter 1, where we considered the fluctuations of accumulated winnings in a game of heads and tails. If a plot of 5000 tosses is blown up by a factor of 10, so that only 500 tosses now come into view, a similar pattern of fluctuations appears at this new scale, and the cluster of points now shows additional clusters within them that were not visible before. Clusters are actually clusters within clusters, down to the level of a single toss (Figure 5.1).

The self-similarity of fractal structures implies that there is some redundancy because of the repetition of details at all

What Is Random?

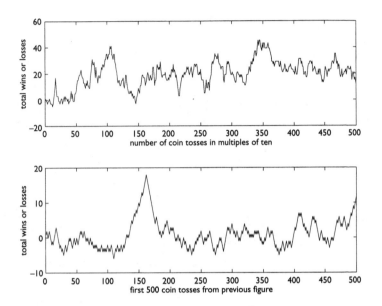

Figure 5.1 Fluctuation in total winning (or losses) in a game of heads and tails with 4000 tosses, plotted every tenth value (upper half), and a closeup of the same figure from 1 to 500 (lower half). The same amount of detail is visible in the plot magnified 10 times as there is in the original.

scales. Even though some of these structures may appear to teeter on the edge of randomness, they actually represent complex systems at the interface of order and disorder. The reason for discussing fractals here is that as you will soon see, there is an identifiable signature of self-similarity that allows us to

The Edge of Randomness

make a connection to the notions of self-organization of complex systems discussed in the previous section.

Fractal structures appear spatially and temporally, but the present discussion is limited to observations of natural phenomena that take place over time, focusing on some measure of the fluctuations to indicate the dynamic interactions that occur among entities of some large system. An example is the dynamics of plankton species in the ocean, as measured by cell counts per milliliter of ocean water sampled hourly and averaged daily, at a mooring off the continental shelf of the United States (Figure 5.2). Reference [1] provides more details.

In general, there will be some measured quantity whose fluctuations vary with the passage of time, such as the concentration of plankton cells in the previous example. I call such a quantity a *fractal signal,* or *fractal time series,* if it exhibits self-similarity. This must be understood in a statistical sense, meaning that *features are similar but not necessarily identical over a large range of scales;* what one sees over smaller portions of the time record is more or less replicated over the entire time series. In all instances it is understood that series represent data obtained from a sequence of measurements of some physical phenomenon.

Early in the nineteenth century the mathematician Joseph Fourier established that any continuous signal of finite duration can be represented as a superposition of overlapping sinusoidal waves, periodic oscillations of different frequencies and amplitudes. The *frequency,* denoted by f, is the reciprocal of the length of the *period,* namely the duration $1/f$ of a complete

What Is Random?

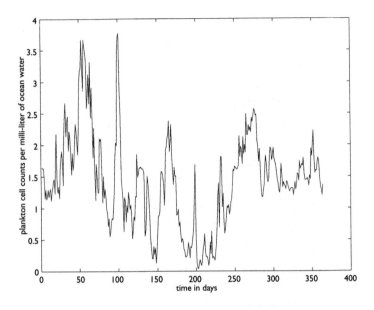

Figure 5.2 Plankton cell counts per milliliter of ocean water as a daily average, over a nearly one-year span. The data were obtained at a mooring off the continental shelf of the east coast of the United States.

cycle, and it measures how many periodic cycles there are per unit time. In Figure 5.3, for example, one sees a portion of two periodic sinusoidal signals (solid lines) having different amplitudes of oscillation. The smaller fluctuation has a period of 10 seconds, and the larger fluctuation has a period of 20 seconds. The frequencies are therefore 0.10 and 0.05 cycles per

The Edge of Randomness

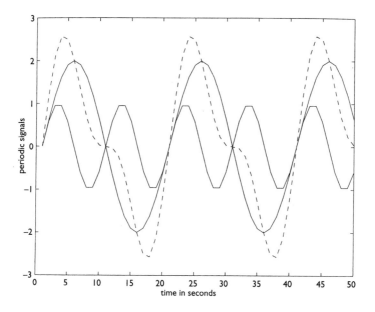

Figure 5.3 Two periodic signals and their superposition (dotted line). They have frequencies of 0.10 (smaller oscillation) and 0.05 cycles (larger oscillation) per second.

second, respectively. Superimposed on the the two oscillations is their sum (dashed line).

Associated with each signal is its *spectrum,* which is a measure of how much variability the signal exhibits corresponding to each of its periodic components. In the usual treatment, the spectrum is expressed as the square of the magnitude of the oscillation at each frequency; it indicates the extent to which the

What Is Random?

magnitudes of separate periodic oscillations contribute to the total signal. If the signal is periodic to begin with, with period $1/f$, then its spectrum is everywhere zero except for a single contribution at the isolated value f, and in the case of a signal that is a finite sum of periodic oscillations the spectrum will exhibit a finite number of contributions at the frequencies of the given oscillations that make up the signal (the spectrum of the dashed curve in Figure 5.3 consists of two isolated values at the frequencies 0.05 and 0.10).

At the other extreme is a signal, called *white noise*, whose sampled values are statistically independent and uncorrelated. It has contributions from oscillations whose amplitudes are uniform over a wide range of frequencies. In this case the spectrum has a constant value, flat throughout the frequency range, and there is no way to distinguish between the contribution of one periodic component and that of any other.

These extremes of signals are reminiscent of the binary strings considered in Chapter 1, in which the indefinite repetition of a deterministic pattern like 01010101 . . . resembles a purely periodic signal, while a random string generated from a Bernoulli ½-process is akin to white noise.

Our interest lies, however, with complex series of data that conform to neither of these opposites, consisting of many superimposed oscillations at different frequencies and amplitudes, with a spectrum that is *approximately* proportional to $1/f^b$ for some number b greater than zero; namely, the spectrum varies inversely with the frequency. These are generically

The Edge of Randomness

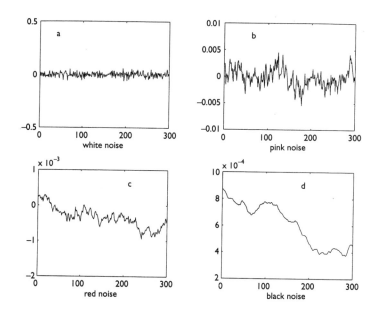

Figure 5.4 Four examples of data signals having approximately a $1/f^b$ spectrum for $b = 0$ (white noise; a), $b = 1$ (pink noise; b), $b = 2$ (red noise; c) and $b = 3$ (black noise; d). Note that the signals get progressively smoother as b increases (very antipersistent to very persistent).

called *1/f noises*. Figure 5.4 illustrates a few cases of signals with spectra having $b = 1$ (called "pink noise"), $b = 2$, and $b = 3$. The range between 1 and 3 often is designated as "red noise," and the term "black noise" is reserved for $b = 3$. For purposes of comparison the figure includes $b = 0$ (approximately flat

What Is Random?

spectrum, or "white noise"). Notice that the degree of irregularity in the signals decreases as b gets larger.

A key observation is that except for $b = 2$, the corresponding fractal series possesses a long-term memory of the past in the sense that fluctuations within any time interval are statistically correlated with changes taking place in a nonoverlapping interval preceeding it, regardless of how long or short the interval. This implies a kind of redundancy due to the near replication of features in the data. Therefore, even though fractal series display considerable variability, there is a core of order buried within them.

For b greater than 2 the correlations are *persistent* in that upward or downward trends tend to maintain themselves. A large excursion in one time interval is likely to be followed by another large excursion in the next time interval of the same length, and this is true no matter how long the interval is. This is sometimes referred to as a "Joseph effect," after the biblical story of seven years of plenty followed by seven years of famine. An example is provided by data concerning the annual discharge levels of the Nile River, compiled by hydrologist Edwin Hurst, which has been shown to be a persistent time series.

With b less than 2 the correlations are *antipersistent* in the sense that an upswing now is likely to be followed shortly by a downturn, and vice versa. This explains why, as you move from antipersistent to persistent, namely as b increases, the curve traced by the sequential data in the time series becomes increasingly less jagged. These assertions can be justified mathematically, but since our purpose here is merely to give an in-

The Edge of Randomness

tuitive overview of the salient features of such temporal records, I sidestep the inessential details.

The plankton data of Figure 5.2 are an example of an antipersistent series, and its spectrum can be roughly approximated by the line $1/f^{5/3}$. Although the plankton data appear haphazard, they are not truly random. In fact, by applying a modification of the ApEn test of Chapter 25.1 we find that ApEn(1) is 4.50, while ApEn(2) is 1.13, which shows that the time series does indeed harbor some redundancy.

Since the spectrum gets progressively smaller as frequency increases, it is intuitively clear that large-amplitude fluctuations are associated with long-wavelength (low-frequency) oscillations, and smaller fluctuations correspond to short-wavelength (high-frequency) cycles. Of particular interest are the pink noises (with b roughly equal to 1), since these most resemble the paradigm of processes balanced between order and disorder. For pink noise the fraction of total variability in the data between two frequencies f_1 and f_2, with f_2 greater than f_1, equals the percentage variability within the interval cf_1 and cf_2 for any positive constant c.[5.2] For instance, up to a multiplicative factor of 100, the percentage of total variability in the wide-frequency range between 100 and 200 cycles per second equals the percentage of variability in the more restricted low-frequency band within 1 and 2 cycles per second. Therefore, there must be fewer large-magnitude fluctuations at lower frequencies than there are small-magnitude oscillations at high frequencies. As the time series increases in length, more and more low-frequency but high-magnitude events are uncovered

What Is Random?

because cycles of longer periods are included, with the longest cycles having periods comparable to the duration of the sampled data. Like the coastline of Britain, small changes are superimposed on larger changes, and so on, to forge a self-similar structure at all scales.

A special role is assigned to $1/f$ spectra, which appears as "flicker noise" in electronics. The engineer Stephan Machlup distinguishes the uniform sound of white noise, "the shshsh sound of running water," from a pink (flicker) sound "pshsh, ktshsh, pok, kshsh . . . in which you can hear the individual events, and big events are less frequent than little events." The high-frequency occurrences are hardly noticed, but the large-magnitude events grab our attention. That is why Machlup titled his paper "Earthquakes, Thunderstorms, and Other $1/f$ Noises."

IBM scientists Richard Voss and John Clarke showed that the fluctuations of loudness as well as the intervals between successive notes in the music of Johann Sebastian Bach have a $1/f$ spectrum, roughly pink. They also artificially created music by varying the frequency and duration of notes using various noise sources. White-noise music, whose successive values are uncorrelated in time, is perceived as stuctureless and irritating, while music from sources that approach black noise seems too predictable and boring because the persistent signal depends strongly on past values. Though one kind of "music" is too unpredictable and the other too structured for either to be interesting, music from a pink noise source sounds just right, a nice balance between surprise and inevitability.

The Edge of Randomness

Physicist Per Bak, together with his colleagues, has devised a demonstrable model of how flicker noises can occur. He considers a sand pile built up by dropping grains of sand, one at a time, onto a flat surface. As the pile increases in size it resembles an inverted cone. When the slope of the cone becomes steep enough, the grains may occasionally slide down, causing a small disruption in the slope, and this realigns the surface somewhat, forcing some grains near the bottom to move away from the pile's edge. Eventually, any additional grains tend to balance the ones leaving the bottom edge, and the system is now at a critical state where any new grain can cause an avalanche of any size along the cone's slope. The spectrum of avalanche sizes at the critical state corresponds to a $1/f$ noise. Since this self-similarity is a robust property of the critical state toward which the sand pile has evolved, I now refer to it, somewhat loosely, as a *$1/f$ law*.

Perturbations in the slope of the pile are contingent events that are correlated in space and time, and future disruptions are influenced by the pile's past history. Bak argues that the sand pile is a paradigm of many natural processes that organize themselves to a critical state at which many small events can trigger a large avalanche, without any ad hoc explanation for the occurrence having to be invoked. One can cite specific data regarding *earthquakes, meteor impacts, stock market fluctuations, atmospheric and oceanic turbulence, forest fires, and species extinctions* as specific systems that scale as $1/f$ laws and for which catastrophic but sporadic events can be expected as naturally as one accepts the hardly noticeable but much more ubiquitous

What Is Random?

benign events that cause little repercussion. This is not unrelated to Kauffman's idea that organization emerges from the unpredictable encounters of individual elements until the system of many entities becomes poised at the edge of randomness, and it finds resonance in Prigogine's argument that organization of complex systems takes place far from a state of equilibrium and order. Many small chance encounters in the past conspire to unleash large disruptions from time to time, unexpected events that punctuate periods of relative quiescence. The hidden order that is latent in the seemingly random data string emerges when we look at it through a suitable prism, namely the spectrum, since this makes apparent scale invariance and redundancy by unveiling that a $1/f$ law had been lurking in the background all along.

Using $1/f$ as a statistical anchor to organize a mass of seemingly unrelated phenomena finds an echo in the preceding two centuries when an attempt was first made to tame chance by appealing to the "law of errors," the Gaussian law. There are some striking differences, however. The Gaussian bell-shaped curve describes the distribution of superpositions of unrelated events, such as the sample average, in which independent zero or one outcomes are added, and these totals spread themselves about the population average, with most near the average and fewer of them further out. By contrast, fractal phenomena hinge on contingency.

If one employs a logarithmic scale on both the horizontal and vertical axes, then any $1/f$ spectrum will appear as roughly a straight line sloping downward from left to right.[5.2] On this

The Edge of Randomness

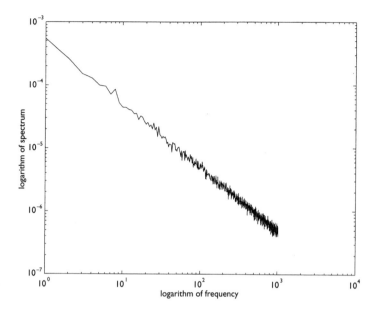

Figure 5.5 The spectrum of a pink noise signal (Figure 5.4 b) versus frequency, plotted on a logarithmic scale.

scale the decrease in volatility from large- to small-magnitude events, as the frequency increases, becomes an instantly recognizable signature of how the magnitude of fluctuations caused by the confluence of many contingent events distribute themselves by a rule of self-similarity. Figure 5.5 exhibits the spectrum of on actual $1/f$ noise.

Data that conform to $1/f$ laws are decidedly non-Gaussian in several other respects. Because of self-similarity there are many more small fluctuations than very large ones, and so the average fluctuation will depend on the ratio of large to small components. The sample average will tend to increase without bound or shrink to zero as more and more pieces of data are included, and the Law of Large Numbers, which held data in its grip in the preceeding chapters, is no longer operable. Moreover, the spread about the population average in the Gaussian case is now ill defined, since dispersion increases as the time series is extended due to the fact that ever larger fluctuations get included; small cycles are added to larger cycles, which are added to even larger ones.

Examples of fractal data sets that lead to $1/f$ spectra abound in natural phenomena, as I have indicated, and are also seen in a variety of physiological contexts, such as heartbeat variability and the electrical activity of neurons. Much as the Gaussian law establishes a form of order in its domain, the $1/f$ laws provide a signature organizing principle for a different category of events. The appearance of a $1/f$ footprint indicates a complex process organized at the brink of randomness. The enthusiasm that the Gaussian law provoked in Francis Galton (recall his unbridled remarks in Chapter 1) is countered today by the following equally unfettered comment of physicist Per Bak: "Self-organized criticality is a law of nature from which there is no dispensation."

Near-random processes, endowed as they are with shifting patterns of order, may conceivably be the consequence of relatively short codes that provide rules of replication and redundancy as a consequence of statistical self-similarity over a wide range of temporal and spatial scales. The rules at one scale beget similar rules at other scales that lead to the emergence of ever more complex structures. This is parodied by Jonathan Swift's verse in reverse:

> So, naturalists observe, a Flea
> Hath smaller fleas that on him prey;
> And these have smaller fleas to bite'em,
> And so proceed *ad infinitum*.

What Good Is Randomness?

Our romp through the many nuances of randomness in this book comes to a close in the present chapter by proposing that you think of uncertainty as something more than an elusive nuisance or an oddity of mathematics.

In antiquity randomness was seen as a cause of disarray and misfortune, disrupting what little order individuals and society had managed to carve out from their surroundings. The world today still contends with uncertainty and risk in the form of natural disasters, acts of terrorism, market downturns, and other outrageous slings of fortune, but there is also a more uplifting view of randomness as the catalyst for life-enhancing changes. Uncertainty is a welcome source of innovation and diversity, providing the raw material for the evolution and renewal of life. Because stasis leads to decay and death, chance fluctuations may guarantee the viability of organisms and the resilience of ecosystems. And, just possibly, chance offers a

What Is Random?

temporary refuge from the inexorable Second Law of Thermodynamics by permitting complex structures to emerge through the exploitation of fortuitous accidents. Finally, chance events relieve the tedium of routine and furnish the element of surprise that makes languages, the arts, and human affairs in general a source of endless fascination.

Randomness, in a nutshell, is what gives meaning to that popular refrain from the Victor Herbert operetta *Naughty Marietta:* "Ah! sweet mystery of life, at last I found thee, Ah! I know at last the secret of it all."

Sources and Further Readings

Chapter 1: The Taming of Chance

The introduction to probability and statistics in this chapter is thoroughly elementary and intuitive. Among the many elementary accounts of statistics, a particularly clear treatment is provided in *Statistics,* a book by Freedman, Pisani, Purves, and Adhikari [29]. To my mind the most authoritative and provocative book on probability remains the classic work *An Introduction to Probability Theory and its Applications* of William Feller [27], where among other topics one will find proofs of Bernoulli's Law of Large Numbers and de Moivre's theorem regarding the normal, or Gaussian, curve. Although Feller's book is very engaging, it is not always easy reading for the beginner.

Sources and Further Readings

There are two delightful articles by Mark Kac called "What Is Random?" [37] that prompted the title of this book, and "Probability" [38]. There are several additional articles on probability in the collection *The World of Mathematics,* edited by James Newman some years ago [52]. I especially recommend the passages translated from the works of James Bernoulli, Pierre Simon Laplace (from which I also cribbed a quotation), and Jules-Henri Poincaré. The Francis Galton quotation also comes from Newman's collection, in an essay by Leonard Tippett called "Sampling and Sampling Error," and Laplace's words are from a translation of his 1814 work on probability in the same Newman set. A philosophical perspective on probability, called simply "Chance," is by Alfred J. Ayer [3].

Technical accounts of statistical tests of randomness, including the runs test, are in the second volume of Donald Knuth's formidable magnum opus *Seminumerical Algorithms* [42].

The idea of chance in antiquity is documented in the book *Gods, Games, and Gambling* by Florence David [22], while the rise and fall of the goddess Fortuna is recounted in a book by Howard Patch [54], from which I took a quotation. Oystein Ore has written an engaging biography of Girolamo Cardano, including a translation of *Liber de Ludo Aleae,* in *Cardano, the Gambling Scholar* [53]. A spirited and very readable account of how the ancients conceptualized the idea of randomness and how this led to the origins of probability is provided in Deborah Bennett's attractive book called *Randomness* [8].

The early history of the rise of statistics in the service of government bureaucracies during the eighteenth and nineteenth centuries is fascinatingly told by Ian Hacking [35] in a book whose title I shamelessly adopted for the present chapter. I also recommend the entertaining book *Against the Gods,* by Peter Bernstein [12], that takes particular aim at the notion of chance and risk in the world of finance.

The perception of randomness as studied by psychologists is summarized in an article by Maya Bar-Hillel and Willem Wagenaar [6]. More will be said about this topic in the next chapter.

Chapter 2: Uncertainty and Information

The story of entropy and information and the importance of coding is the subject matter of information theory. The original paper by Claude Shannon is tough going, but there is a nice expository article by Warren Weaver that comes bundled with a reprint of Shannon's contribution in the volume *The Mathematical Theory of Communication* [62]. There are several other reasonably elementary expositions that discuss a number of the topics covered in this chapter, including those of Robert Lucky [45] and John Pierce [56]. An up-to-date but quite technical version is Thomas Cover and Joy Thomas's *Elements of Information Theory* [20]. A fascinating account of secret codes is David Kahn's *The Code-Breakers* [39]. There are readable accounts of the investigations of psychologists in the perception of randomness by Ruma Falk and Clifford Konold [26] and Daniel Kahneman and Amos Tversky [40], in addition to the previously cited paper by Maya Bar-Hillel and Willem Wagenaar [6].

Sources and Further Readings

Approximate entropy is discussed in an engagingly simple way by Fred Attneave in a short monograph called *Applications of Information Theory to Psychology* [2]. Extensions and ramifications of this idea, as well as the term ApEn, can be found in "Randomness and Degrees of Irregularity" by Steve Pincus and Burton Singer [57] and "Not All (Possibly) 'Random' Sequences Are Created Equal" by Steve Pincus and Rudolf Kalman [58]. Among other things, these authors do not restrict themselves to binary sequences, and they permit arbitrary source alphabets. An elementary account of this more recent work in John Casti's "Truly, Madly, Randomly" [14].

Chapter 3: Janus-Faced Randomness

The Janus algorithm was introduced in Bartlett's paper "Chance or Chaos" [7]. A penetrating discussion of the mod 1 iterates and chaos is available in Joseph Ford's "How Random Is a Coin Toss?" [28].

Random number generators are given a high-level treatment in Donald Knuth's book [42], but this is recommended only to specialists. A more accessible account is in Ivar Ekeland's charming book *The Broken Dice* [25].

The quotation from George Marsaglia is the title of his paper [49].

Maxwell's and Szilard's demons are reviewed in two papers having the same title, "Maxwell's Demon," by Edward Daub [21] and W. Ehrenberg [24]. There is also a most informative paper by Charles Bennett called "Demons, Engines, and the

Sources and Further Readings

Second Law" [9]. The slim volume *Chance and Chaos,* by David Ruelle [60], provides a simplified introduction to many of the same topics covered here and in other portions of our book.

George Johnson's fast-paced and provocative book *Fire in the Mind* [36] also covers many of the topics in this and the next two chapters, and is especially recommended for its compelling probe of the interface between science and faith.

Chapter 4: Algorithms, Information, and Chance

I have glossed over some subtleties in the discussion of algorithmic complexity, but the dedicated reader can always refer to the formidable tome by Li and Vitanyi [44], which also includes extensive references to the founders of this field several decades ago, including the American engineer Ray Solomonoff, who, some believe, may have launched the entire inquiry.

There is an excellent introduction to Turing machines and algorithmic randomness in "What Is a Computation," by Martin Davis [23], in which he refers to Turing–Post programs because of the related work of the logician Emil Post. An even more detailed and compelling account is by Roger Penrose in his book *Shadows of the Mind* [55], and my proof that the halting problem is undecidable in the technical notes is taken almost verbatim from this source.

The best introduction to Gregory Chaitin's work is by Chaitin himself in an article called "Randomness and Mathematical Proof" [16] as well as the slightly more technical

Sources and Further Readings

"Information-Theoretic Computational Complexity" [17]. Chaitin's halting probability Ω is described in an IBM report by Charles Bennett called "On Random and Hard-to-Describe Numbers" [11], which is engagingly summarized by Martin Gardner in "The Random Number Ω Bids Fair to Hold the Mysteries of the Universe" [31]. The Bennett quotation is from these sources.

The distinction between cause and chance as generators of a string of data is explored more fully in the book by Li and Vitanyi mentioned above.

The words of Laplace are from a translation of his 1814 *Essai philosophique sur les probabilités,* but the relevant passages come from the excerpt by Laplace in Newman's *World of Mathematics* [52]. The Heisenberg quotation is from the article "Beauty and the Quest for Beauty in Science," by S. Chandrasekhar [18].

The work of W.H. Zurek that was alluded to is summarized in his paper "Algorithmic Randomness and Physical Entropy" [67].

Quotations from the writing of Jorge Luis Borges were extracted from his essays "Partial Magic in the Quixote" and "The Library of Babel," both contained in *Labyrinths* [13].

Chapter 5: The Edge of Randomness

Charles Bennett's work on logical depth is found in a fairly technical report [10]. A sweeping overview of many of the topics in this section, including the contributions of Bennett,

Sources and Further Readings

is *The Quark and the Jaguar,* by Murray Gell-Mann [32]. Jacques Monod's influential book is called *Chance and Necessity* [51]. Robert Frost's arresting phrase is from his essay "The Figure a Poem Makes" [30], and Jonathan Swift's catchy verse is from his "On Poetry, a Rhapsody" and is quoted in Mandelbrot's book [47].

The quotations from Stephen Jay Gould come from several of his essays in the collection *Eight Little Piggies* [34]. His view on the role of contingency in the evolutionary history of living things is summarized in the article "The Evolution of Life on the Earth" [33]. The other quotations come from the ardently written books *How Nature Works* [4], by Per Bak, and *At Home in the Universe,* by Stuart Kauffman [41]. Ilya Prigogine's thoughts on nonequilibrium phenomena are contained in the celebrated *Order Out of Chaos,* which he wrote with Isabelle Stengers [59]. Cawelti's book is *Adventure, Mystery, and Romance* [15]. A useful resume of the work of Watts and Strogatz is in "It's a Small World" by James Collins and Carson Chow [19].

Benoit Mandelbrot's astonishing book is called *The Fractal Geometry of Nature* [47]. Some of the mathematical details that I left unresolved in the text are provided in a rather technical paper by Mandelbrot and Van Ness [48].

Other, more accessible, sources to self-similarly, $1/f$ laws, and details about specific examples are noted in Bak's book mentioned above; the article "Self-organized Criticality," by Per Bak, Kan Chen, and Kurt Wiesenfeld [5]; Stephan Machlup's "Earthquakes, Thunderstorms, and Other $1/f$

Sources and Further Readings

Noises" [46]; "The Noise in Natural Phenomena," by Bruce West and Michael Shlesinger [65]; "Physiology in Fractal Dimensions," by Bruce West and Ary Goldberger [66]; and the books *Fractals and Chaos,* by Larry Leibovitch [43], and Manfred Schroeder's *Fractals, Chaos, and Power Laws* [61]. All of the quotations come from one or the other of these sources.

Music as pink noise is discussed in "$1/f$ Noise in Music; Music from $1/f$ Noise," by Richard Voss and John Clarke [64]. The plankton data are from [1].

Technical Notes

The following notes are somewhat more technical in nature than the rest of the book and are intended for those readers with a mathematical bent who desire more details than were provided in the main text. The notes are strictly optional, however, and in no way essential for following the the main thread of the discussion in the preceeding chapters.

Chapter 1: The Taming of Chance

1.1.

The material under this heading is optional and is intended to give the general reader some specific examples to illustrate the probability concepts that were introduced in the first chapter.

Technical Notes

In three tosses of a coin, the sample space S consists of 8 possible outcomes HHH, HHT, HTH, THH, THT, TTH, HTT, TTT, with H for head and T for tail. The triplets define mutually exclusive events, and since they are all equally likely, each is assigned probability $1/8$. The composite event "two heads in three tosses" is the subset of S consisting of HHT, HTH, THH (with probability $3/8$), and this is not mutually exclusive of the event "*at least* two heads in three tosses" since the latter event consists of the three triplets HHT, HTH, THH plus HHH (with probability $4/8 = 1/2$).

In the example above, HHT and HTH are mutually exclusive events, and the probability of one or the other happening is $1/8 + 1/8 = 1/4$. By partitioning S into a bunch of exclusive events, the sum of the probabilities of each of these disjoint sets of possibilities must be unity, since one of them is certain to occur.

The independence of events A and B is expressed mathematically by stipulating that the event "both A and B take place" is the product of the separate probabilities of A and B. Thus, for example, in three tosses of a fair coin, the events "first two tosses are heads" and "the third toss is head," which are denoted respectively by A and B, are intuitively independent, since there is a fifty-fifty chance of getting a head on any given toss regardless of how many heads preceded it. In fact, since A consists of HHH and HHT, while B is HHH, HTH, THH, TTH, it follows that the probability of $A = 2/8 = 1/4$ and the probability of $B = 4/8 = 1/2$; however, the probability that "both A and B takes place" is $1/8$, since this event consists only of

HHH. Observe that the product of the probabilities of A and B is ¼ times ½, namely ⅛.

One can also speak of events conditional on some other event having taken place. When the event F is known to have occurred, how does one express the probability of some other event E in the face of this information about F? To take a specific case, let us return to the sample space of three coin tosses considered in the previous paragraph and let F represent "the first toss is a head." Event F has probability ½, as a quick scan of the 8 possible outcomes shows. Assuming that F is true reduces the sample space S to only four elementary events HTT, HTH, HHT, HHH. Now let E denote "two heads in 3 tosses." The probability of E equals ⅜ if F is not taken into account. However, the probability of E given that F has in fact taken place is ½, since the number of possibilities has dwindled to the two cases HHT and HTH out of the four in F. The same kind of reasoning applies to any finite sample space and leads to the *conditional probability* of E given F. Note that if E and F are independent, then the conditional probability is simply the probability of E, because a knowledge of F is irrelevant to the happening or nonhappening of E.

1.2.

This technical aside elaborates on the interpretation of infinite sample spaces of unlimited Bernoulli ½-trials that began with the discussion of the Strong Law of Large Numbers. Though not necessary for what follows in the book, it offers the mathematically inclined reader some additional insights

Technical Notes

into the underpinnings of modern probability theory. I will use some mathematical notions that may already be familiar to you, but in any case I recommend that you consult Appendices A and B, which cover the required background material in more detail.

The set of all numbers in the unit interval between zero and one can be identified in a one-to-one manner with the set of all possible nonterminating strings of zeros and ones, provided that strings with an unending succession of ones are avoided. The connection is that any x in the unit interval can be written as a sum

$$x = a_1/2 + a_2/2^2 + a_3/2^3 \ldots,$$

where each a_k, $k = 1, 2 \ldots$, denotes either 0 or 1. For example, $3/8 = 0/2^1 + 1/2^2 + 1/2^3$, with the remaining terms all zero. The number x is now identified with the infinite binary string $a_1 a_2 a_3 \ldots$. With this convention, $x = 3/8$ corresponds to the string 011 00.... Similarly, $1/3 = 01010101\ldots$. In some cases two representations are possible, such as $1/4 = 01000\ldots$ and $1/4 = 00111\ldots$, and when this happens we opt for the first possibility. So far, this paraphrases Appendix B, but now comes the connection to probability.

Each unending sequence of Bernoulli ½-trials gives rise to an infinite binary string and therefore to some number x in the unit interval. Let S be the sample space consisting of all possible strings generated by this unending random process. Then an event E in S is a subset of such strings, and the *probability of E is taken to be the length of the subset of the unit interval that corre-*

Technical Notes

sponds to E. Any two subsets of identical length correspond to events having the same probability.

If the subset consists of nonoverlapping intervals within the unit interval, then its length is the sum of the lengths of the individual subintervals. For more complicated subsets, the notion of length becomes problematic, and it is necessary to invoke the concept of measure (see the discussion below, later in this note).

Conversely, any x in the unit interval gives rise to a string generated by a Bernoulli ½-process. In fact, if x corresponds to $a_1 a_2 a_3 \ldots$, then $a_1 = 0$ if and only if x belongs to the subset [0, ½), which has length ½. This means that the probability of a_1 being zero is ½; the same applies if $a_1 = 1$. Now consider a_2. This digit is 0 if and only if x belongs to either [0, ¼) or [½, ¾). The total *length* of these disjoint intervals is ½, and since they define mutually exclusive events, the probability of a_2 being zero is necessarily ½. Obviously, the same is true if $a_2 = 1$. Proceeding in this fashion it becomes evident that each a_k, for k = 1, 2, . . . , has an equal probability of being zero or one. Note that regardless of the value taken on by a_1, the probability of a_2 remains ½, and so a_1 and a_2 are independent. The same statistical independence applies to any succession of these digits. It follows that the *collection of numbers in the unit interval is identified in a one-to-one fashion with all possible outcomes of Bernoulli ½-trials.*

If s is a binary string of length n, let Γ_s denote the set of all infinite strings that begin with s. Now, s itself corresponds to some terminating fraction x, and since the first n positions of

Technical Notes

the infinite string have already been specified, the remaining digits sum to a number whose length is at most $½^n$. To see this, first note that any sum of terms of the form $a_k/2^k$ can never exceed in value a similar sum of terms $½^k$, since a_k is always less than or equal to one. Therefore the infinite sum of terms $a_k/2^k$ starting from $k = n + 1$ is at most equal to the corresponding sum of terms $½^k$, and as Appendix A demonstrates, this last sum is simply $½^n$.

Therefore, the event Γ_s corresponds to the interval between x and $x + ½^n$, and this interval has length $½^n$. It follows that the event Γ_s has probability $½^n$. Thus, if E is the event "all strings whose first three digits are one," then E consists of all numbers that begin with $½ + ¼ + ⅛ = ⅞$, and this gives rise to the interval between $⅞$ and 1, whose length is $⅛$. Therefore the probability of E is $⅛$. If you think of the Γ_s as elementary events in S, then they are uniformly distributed, since their probabilities are all equal.

There is a technical complication here regarding what is meant by the length of a subset when the event E is so contrived that it does not correspond to an interval in the ordinary sense. However, a resolution of this issue would get us tied up in the nuances of what is known in mathematics as *measure theory,* a topic that need not detain us. For our purposes it is enough to give a single illustration to convey the essence of what is involved. Let E denote "the fourth trial is a zero." This event is realized as the set all strings whose first three digits are one of the $2^3 = 8$ possibilities 111, 110, . . . , 000 and whose fourth digit is zero. Each of these possibilities corresponds to a

Technical Notes

number that begins in one of 8 distinct locations in the unit interval and extends for length $1/16$ to the right. For example, 1010 represents the number $5/8$, and the remaining sequence sums to at most another $1/16$. It follows that E is the union of 8 disjoint events, each of length $1/16$, summing to $1/2$, and this is the probability of E (again, a more detailed discussion can be found in Appendix A).

An unexpected feature of this infinite sample space is that since probability zero must correspond to a set of zero "length," called *measure zero,* it is possible that an event of zero probability may itself correspond to an infinite collection of strings! To see this we consider the subset R of all rational numbers (namely, common fractions) in the unit interval. These fractions can be written as an infinite sequence of infinite sequences in which the numerator is first 1, then 2, and so on:

$$1/2, 1/3, 1/4, \ldots$$
$$2/2, 2/3, 2/4, \ldots$$
$$3/3, 3/4, 3/5, \ldots$$
$$\ldots$$
$$\ldots$$

With a zigzag tagging procedure these multiple arrays of fractions can be coalesced into a single infinite sequence of rational numbers, as described in the ninth chapter of Ian Stewart's introductory book [63]. Omitting repetitions, the first few elements of this sequence are $1/2, 1/3, 2/3, 2/3, 1/4, 1/5, 3/4, \ldots$

Let ϵ be an arbitrarily small positive number and enclose the nth element of the sequence within an interval of length ϵ

times $\frac{1}{2}^n$ for $n = 1, 2, \ldots$. Then R is totally enclosed within a set S whose total length cannot exceed ϵ times the sum $\frac{1}{2} + \frac{1}{2}^2 + \frac{1}{2}^3 + \ldots$, which (Appendix A again!) is simply ϵ. But ϵ was chosen to be arbitrarily small, and therefore the "length," or measure, of S must necessarily be zero. The probability of "a string corresponds to a rational number" is consequently zero. In the language of measure theory it is often stated that an event has probability 1 if it holds "almost everywhere," that is, except for a negligible subset of measure zero.

For balanced coins ($p = \frac{1}{2}$), the content of the Strong Law of Large Numbers is that S_n/n differs from $\frac{1}{2}$ by an arbitrarily small amount for all large enough n, *except* for a subset of unlimited Bernoulli $\frac{1}{2}$-trials having measure zero. This is the caveat that I mentioned earlier, formulated now in the language of measure theory.

I can now restate the assertion made in the text about normal numbers by saying that *all numbers are normal except for a set of measure zero.*

1.3.

The simple MATLAB *m*-file below is for those readers with access to MATLAB who wish to generate Bernoulli *p*-trials for their own amusement:

```
% CODE TO GENERATE SEQUENCES
% FROM A BERNOULLI P-PROCESS. The output
% is a binary string of length n.
```

Technical Notes

```
n = input('string size is . . .');
p = input('probability of a success is . . .');
ss = [];
for k = 1: n
if rand < p
s = 1;
else s = 0; end
ss = [ss, s];
end
ss
```

Chapter 2: Uncertainty and Information

2.1.

The following comments are intended to clarify the discussion of message compression when there are redundancies.

The Law of Large numbers discussed in the previous chapter asserts that in a long enough sequence of n Bernoulli p-trials with a likelihood p of a success (namely a one) on any trial, the chance that the proportion of successes S_n/n deviates from p by an arbitrarily small amount is close to unity. What this says, in effect, is that in a long binary string, it is very likely that the number of 1's, which is S_n, will nearly equal np, and of course, the number of 0's will nearly equal $n(1 - p)$. The odds are then overwhelmingly in favor of such a string being close to a string in which the digit one appears np times and the digit zero appears $n(1-p)$ times. This is because successive Bernoulli trials are independent, and as we know, the probability of a particular string of independent events, namely the choice of zero or one, must equal the product of the probabilities of the individual events. The product in question is

$$p^{np}(1-p)^{n(1-p)},$$

and it is equal to $\frac{1}{2}^{nh}$ for some suitable h yet to be determined. Taking logarithms to the base two of both sides (see Appendix C), we obtain

$$np \log p + n(1-p) \log(1-p) = -nh \log 2.$$

One can now solve for h and obtain, noting that $\log 2 = 1$,

$$h = -[p \log p + (1-p) \log(1-p)],$$

which is the entropy H of the source. It is therefore likely that a binary string of length n has very nearly a probability 2^{-nH} of occurring. Stated differently, the fraction of message strings having a probability 2^{-nH} of appearing is nearly one. It follows that there must be altogether about 2^{nH} such message strings, and it is highly improbable that an actual message sequence is not one of them. Even so, these most likely messages constitute a small fraction of the 2^n strings that are possible, since $2^{nH}/2^n$ is tiny for large n. An exception to this is when $p = \frac{1}{2}$ because H equals unity in that case.

As a postscript to this argument I remind you that the Strong Law of Large Numbers asserts that in an infinite sequence of

Technical Notes

Bernoulli p-trials, ones and zeros occur with frequencies p and $1-p$ with probability one. It should be fairly plausible now that "probability one" corresponds to "measure zero," as it was defined in the technical note 1.2, whenever p does not equal ½.

2.2.

This note expands on the Shannon coding scheme.

An alphabet source has k symbols whose probabilities of occurrence $p_1, p_2, \ldots, p_k$ are assumed to be listed in decreasing order. Let P_r denote the sum $p_1 + \ldots + p_{r-1}$, for $r = 1, \ldots, k$, with P_1 set to 0. The binary expansion of P_r is a sum $a_i/2^i$, for $i = 1, 2, \ldots$ (Appendix B). Truncate this infinite sum when i is the first integer, call it l_r, that is no smaller than $-\log p_r$ and, by default, less than $1 - \log p_r$. The binary code for rth symbol is then the string $a_1 a_2 \ldots a_{l_r}$.

For example, for a source alphabet $a, b, c,$ and d with corresponding probabilities ½, ¼, ⅛, and ⅛, $P_1 = 0$, $P_2 = ½ + 0/2^2$, $P_3 = ½ + ½^2 + 0/2^3$, and $P_4 = ½ + ½^2 + ½^3$. Now, $-\log ½ = 1$, $-\log ½^2 = 2$, and $-\log ½^3 = 3$, and so, following the prescription given above, the code words for the source symbols $a, b, c,$ and d are 0, 10, 110, 111, respectively, with lengths 1, 2, 3, and 3.

For $s = r + 1, \ldots, k$ the sum P_s is no less than P_{r+1}, which, in turn, equals $P_r + p_r$. As you saw, $-\log p_r$ is less than or equal to l_r, which is equivalent to stating that p_r is no less than $½^{l_r}$ (see Appendix C for details on manipulating logarithms). Therefore, the binary expansion of P_s differs from that of P_r by at least one digit in the first l_r places, since $½^{l_r}$ is being added in. If the

Technical Notes

last digit of P_r is 0, for instance, adding $1/2^{l_r}$ changes this digit to 1. This shows that each code word differs from the others, and so the k words together are a decodable prefix code. This is certainly true of the illustration above, in which the code is 0, 10, 110, 111.

The Shannon code assigns short words to symbols of high probability and long codes to symbols of low probability. The average code length L is defined to be the sum over r of $l_r p_r$, and the discussion above now readily establishes that L is no smaller than-sum of $p_r \log p_r$ and certainly less than the sum of $(p_r - p_r \log p_r)$. The probabilities p_r add to 1 for $r = 1, \ldots, k$, since the symbols represent mutually exclusive choices. Also, you recognize the sums on the extreme right and left as the entropy H of the source. Putting everything together, therefore, we arrive at the statement that average code length of the Shannon prefix code differs from the entropy, the average information content of the source, by no more than 1. A message string from a source of entropy H can therefore be encoded, on average, with nearly H bits per symbol.

2.3.

In defining ApEn(k) an alternative approach would be to formally compute the *conditional entropy* of a k-gram given that we know its first k-1 entries. This expression, which is not derived here, can be shown to equal $H(k) - H(k-1)$, and this is identical to the definition of ApEn(k) in the text.

Technical Notes

2.4.

A simple program written as a MATLAB m-file implements the ApEn algorithm for *k* up to 4, and is included here for those readers who have access to MATLAB and would like to test their own hand-tailored sequences for randomness.

```
% CODE TO TEST FOR RANDOMNESS
% IN ZERO-ONE STRINGS
% input is digit string u as a column of length n
% and a block length k that is no greater than 4.
%
% the matrix below is used in the computation:
%
A = [1 1 1 1 1 1 1 1 0 0 0 0 0 0 0 0
     1 1 1 1 0 0 0 0 1 1 1 1 0 0 0 0
     1 1 0 0 1 1 0 0 1 1 0 0 1 1 0 0
     1 0 1 0 1 0 1 0 1 0 1 0 1 0 1 0];
%
ApEn = []; h = [];
for m = 1:k
N = n - m + 1;
p = 2^m;
Q = A(1:m, 1:2^(4 - m):16);
U = zeros(N,p);
r = zeros(1,p);
for j = 1:p
for i = 1:N
```

```
U(i,j) = norm(u(i:i + k − 1) − Q(:,j), inf );
end
r( j) = sum(spones(find(U(:,j)==0)));
end
rr = find(r > 0); r = r(rr);
h(m) = log2(N) − sum(r.*log2(r))/N;
end
ApEn(1) = h(1);
% IF k IS GREATER THAN ONE:
for s = 2:k
ApEn(s) = h(s) − h(s − 1);
end
ApEn
```

Chapter 3: Janus-Faced Randomness

Some of the missing details regarding the use of symbolic dynamics are filled in here. Our concern is how to represent the action of the Janus algorithm and its inverse.

3.1.

Begin by writing u_0 as a sum $a_1/2 + a_2/4 + a_3/8 + \ldots$ as described in the text, which implies that $a_1\, a_2\, a_3 \ldots$ represents u_0. Then u_1 is either $u_0/2$ or $½ + u_0/2$. In the first instance the sum becomes $a_1/4 + a_2/8 + a_3/16 + \ldots$, and in the second it is $½ + a_1/4 + a_2/8 + \ldots$. In either case, you obtain u_1 as $b_1/2 + a_1/4 + a_2/8 + \ldots$ with b_1 being zero or one. Therefore, u_1 is represented by the string $b_1\, a_1\, a_2 \ldots$. Similarly, u_2 becomes

Technical Notes

$b_2/2 + b_1/4 + a_2/8 + \ldots$, and so u_2 is identified by the sequence $b_2\, b_1\, a_1\, a_2 \ldots$.

3.2.

By contrast with the previous note, the mod 1 scheme starts with a seed v_0, which can be expressed as $q_1/2 + q_2/4 + \ldots$, and so the binary sequence $q_1\, q_2\, q_3 \ldots$ represents v_0. If v_0 is less than $\frac{1}{2}$, then $q_1 = 0$, and q_1 is 1 when v_0 is greater than or equal to $\frac{1}{2}$. In the first case, $v_1 = 2v_0 \pmod 1$ equals $q_2/2 + q_3/8 + \ldots \pmod 1$, while in the second it is $1 + q_2/2 + q_3/8 + \ldots \pmod 1$. By definition of mod 1 only the fractional part of v_1 is retained, and so 1 is deleted in the last case, while mod 1 is irrelevant in the first because the sum is always less than or equal to 1 (see Appendix A). In either situation the result is the same: v_1 is represented by the sequence $q_2\, q_3\, q_4 \ldots$.

3.3.

It is of interest, in connection with our discussion of pseudorandom number generators, to observe when the seed x_0 of the mod 1 scheme will lead to iterates that repeat after every k iterates, for some $k = 1, 2, \ldots$, a situation that is expressed by saying that *the iterates are periodic with period k*. If, as usual, the binary representation of x_0 is $a_1\, a_2\, a_3 \ldots$, then the mod 1 algorithm lops off the leftmost digit at each step, and so after k steps, the remaining digits are identical to the initial sequence; the block $a_1 \ldots a_k$ of length k is the same as the block $a_{k+1} \ldots a_{2k}$. There are therefore 2^k possible starting "seeds" of period k, one for each of the different binary blocks, and these all lead to *cycles of length* k. For example, 0, ⅓, ⅔, 1 are the $2^2 = 4$ starting points of period 2 and each generates a cycle of length 2. What this means for a period-k sequence is that $2^k x_0$ (mod 1) is identical to x_0, or, as a moment's reflection shows, $(2^k - 1)x_0$ must equal some integer between 0 and $2^k - 1$. It follows that x_0 is one of the fractions $i/(2^k - 1)$, for $i = 0, 1, \ldots, 2^k - 1$. For instance, with $k = 3$, the seed x_0 leads to period-3 iterates (cycles of length 3) whenever it is one of the $2^3 = 8$ starting values 0, ⅐, 2/7, ..., 6/7, 1. If the initial choice is ⅐, the subsequent iterates are 2/7, 4/7, 8/7, with 8/7 identical to ⅐ (mod 1). In terms of binary expansions the periodicity is even more apparent, since the binary representation of ⅐ is 001001001 ..., with 001 repeated indefinitely.

3.4.

There is an entertaining but totally incidental consequence of the consideration in the preceeding note for readers familiar with congruences (an excellent exposition of congruences and modular arithmetic is presented in the third chapter of Ian Stewart's introductory book [63]). Whenever an integer j divides k, the $2j$ cycles of length j are also cycles of length k, since each j-cycle can be repeated k/j times to give a period-k cycle. However, if k is a prime number, the only divisors of k are 1 and k itself, and therefore the 2^k points of period k distibute themselves as distinct clusters of k-cycles. Ignoring the points 0 and 1, which are degenerate cycles having all periods, the remaining $2^k - 2$ points therefore arrange themselves into some integer multiple of cycles of length k. That is, $2^k - 2$ equals

Technical Notes

mk, for some integer m. For example the $2^3 - 2 = 6$ starting values of period 3 other than 0 and 1 group themselves into the two partitions {$1/7$, $2/7$, $4/7$} and {$3/7$, $6/7$, $5/7$}. Now, $2^k - 2 = 2(2^{k-1} - 1)$ is an even number, and since k must be odd (it is prime), it follows that m is even. Therefore, $2^{k-1} - 1$ is a multiple of k, which means that 2^{k-1} equals 1 (mod k). This is known as *Fermat's theorem* in number theory (not Fermat's celebrated last theorem!):

$$2^{k-1} = 1 \text{ (mod } k\text{), whenever } k \text{ is a prime.}$$

3.5.

An argument similar to that in the preceeding note establishes the result of George Marsaglia about random numbers in a plane. Denote two successive iterates of the mod 1 algorithm by $x_k = 2^k x_0$ (mod 1) and $x_{k+1} = 2^{k+1} x_0$ (mod 1), where x_0 is a fraction (a truncated seed value) in lowest form p/q for suitable integers p, q. The pair (x_{k+1}, x_k) is a point in the unit square. Now let a, b denote two numbers for which $a + 2b = 0$ (mod q). If q is 9, for instance, then $a = 7$, $b = 20$ satisfy $a + 2b = 27$, which is a multiple of 9 and therefore equal to zero (mod 9). There are clearly many possible choices for a and b, and there is no loss in assuming that they are integers. It now follows that $ax_k + bx_{k+1} = 2^k x_0 (a + 2b)$ mod 1, and since $a + 2b$ is some multiple m of q, we obtain $ax_k + bx_k = mp2^k$ (mod 1), since the q's cancel. But $mp2^k$ is an integer, and so it must equal zero mod 1. This means that $ax_k + bx_{k+1} = c$, for some integer constant c, which is the equation of a straight line in the plane.

Technical Notes

What this says, in effect, is that the "random number" pair (x_{k+1}, x_k) must lie on one of several possible lines that intersect the unit square. The same argument applies to triplets of points, which can be shown to lie on planes slicing the unit cube, and to k-tuples of points, which must reside on hyperplanes intersecting k-dimensional hypercubes.

3.6.

Among the algorithms that are known to generate deterministic chaos, the most celebrated is the logistic algorithm defined by

$$x_n = r x_n(1 - x_n)$$

for $n = 1, 2, \ldots$ and a seed x_0 in the unit interval. The parameter r ranges in value from 1 to 4, and within that span there is a remarkable array of dynamical behavior. For r less than three, all the iterates tend to a single value as n gets larger. Thereafter, as r creeps upward from three, the successive iterates become more erratic, tending at first to oscillate with cycles that get longer and longer as r increases, and eventually, when r reaches four, there is full-blown chaos. The intimate details of how this all comes to pass have been told and retold in a number of places, and I'll not add to these accounts (see, for example, the classic treatment by Robert May [50]). My interest is simply to mention that in a realm where the iterates are still cycling benignly, with r between 3 and 4, the intrusion of a little bit of randomness jars the logistic algorithm into behaving chaotically. It therefore becomes difficult, if not impossible, to disentangle whether one has a deterministic algorithm corrupted

Technical Notes

by noise or a deterministic algorithm that because of sensitivity to initial conditions only mimics randomness. This dilemma is a sore point in the natural sciences whenever raw data needs to be interpreted either as a signal entwined with noise or simply as a signal so complicated that it only appears noisy.

Chapter 4: Algorithms, Information, and Chance

4.1.

I provide a succinct account of Turing's proof that the halting problem is undecidable. Enumerate all possible Turing programs in increasing order of the size of the numbers represented by the binary strings of these programs and label them as $T_1, T_2, \ldots$. The input data, also a binary string, is similarly represented by some number $m = 1, 2, \ldots$, and the action of T_n on m is denoted by $T_n(m)$. Now suppose that there is an algorithm for deciding when any given Turing program halts when started on some particular input string. More specifically, assume the existence of a program P that halts whenever it has demonstrated that some Turing computation never halts. Since P depends on both n and m, we write it as $P(n, m)$; $P(n, m)$ is alleged to stop whenever $T_n(m)$ does not halt.

If m equals n, in particular, then $P(n,n)$ halts if $T_n(n)$ doesn't halt. Since $P(n,n)$ is a computation that depends only on the input n, it must be realizable as one of the Turing programs, say

Technical Notes

$T_k(n)$, for some integer k, because T_1, T_2, ... encompass all possible Turing machines. Now focus on the value of n that equals k to obtain $P(k,k) = T_k(k)$, from which we assert that if $P(k,k)$ stops, then $T_k(k)$ doesn't stop. But here comes the denouement: $P(k,k)$ equals $T_k(k)$, and so the statement says that *if $T_k(k)$ halts, then $T_k(k)$ doesn't halt!* Therefore, it must be that $P_k(k)$ in fact does not stop, for if it did then it could not, a blatant contradiction. It follows from this that $P(k,k)$, which is identical to $T_k(k)$, also cannot stop, and so the program P is *not able to decide* whether the particular computation $T_k(k)$ halts *even though you know that it does not.*

4.2.

The definition of Ω requires that we sum the probabilities 2^{-l} of all halting programs p of length l, for $l = 1, 2, \ldots$. Let Γ_s correspond to the interval $[s, s + \frac{1}{2}^l)$, where s is understood to be the binary expansion of the fraction corresponding to the finite string s, and l is the length of s. For example, $s = 1101$ is the binary representation of $^{13}/_{16}$, and Γ_s is the interval $[^{13}/_{16}, ^{13}/_{16} + ^{1}/_{16})$, or $[^{13}/_{16}, ^{7}/_{8})$, since $l = 4$ (the intervals Γ_s were introduced in the technical note 1.2 to Chapter 1 in a slightly different manner). If a string s_1 is prefix to string s_2 then Γ_{s_2} is contained within the interval Γ_{s_1}. To see this, observe that s_2 coresponds to a number no less than s_1, while the quantity 2^{-l_2} is no greater than 2^{-l_1}, since the length l_2 of s_2 cannot be less than the length l_1 of s_1.

The halting programs evidently cannot be prefixes of each other, and therefore the intervals Γ_s must be disjoint (cannot

Technical Notes

overlap) from one another. The argument here is that if p and q denote two such programs and if q is not shorter than p, then q can be written as a string s of the same length as p, followed possibly by some additional bits. Since p *is not* a prefix of q, the string s is different from p even though their lengths are the same, and so Γ_s and Γ_p are *different intervals*. However s *is* a prefix of q, and therefore Γ_q is contained within Γ_s, from which it follows that Γ_q and Γ_p are disjoint. Since each of the nonoverlapping intervals Γ resides within the same unit interval $[0, 1]$, their sum cannot exceed the value one. Each halting program of length l now corresponds to one of the disjoint intervals of length 2^{-l}, which establishes that Ω, the sum of the probabilities 2^{-l}, is necessarily a number between zero and one.

4.3.

It was mentioned that $\log n$ bits suffice to code an integer n in binary form. This is shown in Appendix C. Also, since 2^n increases much faster than does n, $\log(2^n/n)$ will eventually exceed $\log 2^m$ for any fixed integer m, provided than n is taken large enough. This means that $n\text{-}\log n$ is greater than m for all sufficiently large n, a fact that is needed in the proof of Chaitin's theorem.

4.4.

It has been pointed out several times that a requirement of randomness is that a binary sequence define a *normal number*. This may be too restrictive a requirement for finite sequences, however, since it implies that zeros and ones should be equidistributed. Assuming that n is even, the number of strings containing exactly $n/2$ ones is approximately proportional to the reciprocal of the square root of n (see the paper by Pincus and Singer [57]). This means that randomness in the strict sense of normality is very unlikely for lengthy but finite strings. The problem is that one is trying to apply a property of infinite sequences to truncated segments of such sequences. This shows that randomness defined strictly in terms of maximum entropy or maximum complexity, each of which imply normality, may be too limiting. The reader may have noticed the more pragmatic approach adopted in the present chapter, in which randomness as maximum complexity is redefined in terms of c-randomness for c small relative to n. With this proviso most long strings are now very nearly random, since the complexity of a string is stipulated to be comparable (but not strictly equal) to its length. The same ruse applies in general to any lengthy but finite block of digits from a Bernoulli ½-process, since the Strong Law of Large Numbers ensures that zeros and ones are very nearly equidistributed with high probability. Again, most long sequences from a Bernoulli ½-process are now random with the proviso of "very nearly." In Chapter 2 the notion of approximate entropy, which measures degrees of randomness, implicitly defines a binary block to be random if its entropy is very nearly maximum. As before, most long strings from Bernoulli ½-trials are now random in this sense.

4.5.

Champernowne's normal number, which will be designated by the letter c, has already been mentioned several times in this

Technical Notes

book, but I cannot resist pointing out another of its surprising properties. The binary representation of c is, as we know, the sequence in which 01 is followed by the four doublets 00 01 10 11, then followed by the eight triplets 000 001 . . . , and so forth. Every block of length k, for any positive integer k, will eventually appear within the sequence. If the inverse of the Janus algorithm, namely the mod 1 algorithm, is applied to c, then after m iterations, its m leftmost digits will have been deleted. This means that if x is any number in the unit interval, the successive iterates of the mod 1 algorithm applied to c will eventually match the first k digits of the binary representation of x; one simply has to choose m large enough for this to occur. Thus c will approximate x as closely as one desires simply by picking k sufficiently large that the remaining digits, from $k + 1$ onward, represent a negligible error. It follows that the iterative scheme will bring Champernowne's number in arbitrary proximity to any x between zero and one: The successive iterates of c appear to jump pell-mell all over the unit interval. Though the iterates of the inverse Janus algorithm follow a precise rule of succession, the actual values seem random to the unitiated eye as it attempts to follow what seems to be a sequence of aimless moves.

Chapter 5: The Edge of Randomness

The discussion in this chapter simplifies the notion of Fourier analysis and spectrum of a time series. The technically prepared reader will indulge any lapses of rigor, since they allow

Technical Notes

me to attain the essence of the matter without getting cluttered in messy details that would detract from the central argument.

5.1.

A version of the ApEn procedure is employed here that is not restricted to binary data strings as in Chapter 2. Suppose that $u_1, u_2, \ldots, u_n$ represent the values of a time series sampled at a succession of n equally spaced times. Pincus and Singer established an extension of ApEn that applies to blocks of data such as this that is similar in spirit to the algorithm presented in Chapter 2. I leave the details to their paper [57].

5.2.

Some of the properties of pink spectra begin with the observation that a spectrum is self-similar. To see this, multiply the frequency f by some constant c greater than one and denote the spectrum by $s(f)$ to indicate its dependence on the frequency. Since $s(f)$ is approximately k/f, for some positive constant of proportionality k, then $s(cf)$ is some constant multiple of $s(f)$. Therefore, $s(cf)$ has the same shape as $s(f)$, which is self-similarity. The integral of the spectrum $s(f)$ between two frequencies f_1 and f_2 represents the total variability in a data series within this range (this is the first and last time in this book that a calculus concept is mentioned!), and from this it is not hard to show that the variability within the frequency band (f_1, f_2) is the same as that within the band (cf_1, cf_2) since the integral of $s(f)$ evaluates to $\log f_2/f_1$, which, of course, doesn't change if the frequencies are multiplied by c.

Using logarithms (Appendix C), it is readily seen that the logarithm of the spectrum is approximately equal to some constant minus the logarithm of the frequency f, which is the equation of a straight line with negative slope. This is the characteristic shape of a $1/f$ spectrum when plotted on a logarithmic scale.

Appendix A: Geometric Sums

The notation S_m is used to denote the finite sum $1 + a + a^2 + a^3 + \ldots + a^m$, where a is any number, m any positive integer, and a^m is the mth power of a. This is called a *geometric sum,* and its value, computed in many introductory mathematics courses, is given by

$$S_m = (1 - a^{m+1})/(1 - a).$$

It is straightforward to see, for example, that if a equals 2, the finite sum S_m is $2^{m+1} - 1$.

When a is some number greater than zero and less than one, the powers a^{m+1} get smaller and smaller as m increases (for example, if $a = 1/3$, then $a^2 = 1/9$, $a^3 = 1/27$, and so forth) and so the expression for S_m tends to a limiting value $1/(1 - a)$ as m

Appendix A: Geometric Sums

increases without bound. In effect, the *infinite sum* $1 + a + a^2 + a^3 + \ldots$ has the value $1/(1 - a)$.

In a number of places throughout the book the infinite sum $a + a^2 + a^3 + \ldots$ will be required, which starts with a rather than 1, and this is 1 less than the value $1/(1 - a)$, namely $a/(1 - a)$.

Of particular interest is the case in which a equals $\frac{1}{2}$, and it is readily seen that $\frac{1}{2} + \frac{1}{2}^2 + \frac{1}{2}^3 + \ldots$ equals $\frac{1}{2}$ divided by $\frac{1}{2}$, namely 1.

There is also a need for the infinite sum that begins with a^{m+1}: $a^{m+1} + a^{m+2} + a^{m+3} + \ldots$. This is evidently the difference between the infinite sum of terms that begins with a, namely $1/(1 - a)$, and the finite sum that terminates with a^m, namely S_m; the difference is $a^{m+1}/(1 - a)$.

The formulas derived so far are summarized below in a more compact form by using the notation Σ_m to indicate the *infinite sum that begins with a^{m+1}* for any $m = 0, 1, 2, \ldots$. This shorthand is not used in the book proper and is introduced here solely as a convenient mnemonic device:

$$\Sigma_m = a^{m+1}/(1 - a);$$
$$\Sigma_0 = a/(1 - a).$$

Here are two examples that actually occur in the book:

Find the finite sum of powers of 2 from 1 to $n - c - 1$ for some integer c smaller than n, namely the sum $2 + 4 + 8 + \ldots + 2^{n-c-1}$; this is 1 less than S_{n-c-1}, with the number a set equal to 2. A simple computation establishes that the term S_{n-c-1}

Appendix A: Geometric Sums

reduces to $2^{n-c} - 1$, and therefore the required sum is $2^{n-c} - 2$. This is used in Chapter 4.

Find the infinite sum of powers of ½ from $m + 1$ onward. This is the expression Σ_m with the number a set to ½, and the first of the formulas above shows that it equals $½^m$ or, as it is sometimes written, 2^{-m}. In the important special case of $m = 0$, namely the sum $½ + ¼ + ⅛ + \ldots$, the total is simply 1.

Appendix B: Binary Notation

Throughout the book there are frequent references to the binary representation of whole numbers, as well as to numbers in general within the interval from 0 to 1. Let us begin with the whole numbers $n = 1, 2, \ldots$.

For each integer n find the largest possible power of 2 smaller than n and call it $2k_1$. For instance, if $n = 19$, then $k_1 = 4$, since 2^4 is less than 19, which is less than 2^5.

The difference between n and $2k_1$ may be denoted by r_1. Now find the largest power of 2, call it $2k_2$, that is smaller than r_1, and let r_2 be the difference between r_1 and $2k_2$. Repeat this procedure until you get a difference of zero, when n is even, or one, when n is odd. It is evident from this finite sequence of steps that n can be written as a sum of powers of 2 (plus 1, if n is odd).

Appendix B: Binary Notation

For example, $20 = 16 + 4 = 2^4 + 2^2$. Another instance is $77 = 64 + 8 + 4 + 1 = 2^6 + 2^3 + 2^2 + 2^0$ (by definition, $a^0 = 1$ for any positive number a). In general, any positive integer n may be written as a finite sum

$$n = a_{m-1} 2^{m-1} + a_{m-2} 2^{m-2} + \ldots + a_1 2 + a_0, (\star)$$

where the m coefficients $a_{m-1}, \ldots, a_0$, are all either 0 or 1 and the leading coefficient a_{m-1} is always 1.

The binary representation of n is the string of digits $a_{m-1} a_{m-2} \ldots a_0$ in the sum ($\star$), and the string itself is often referred to as a binary string.

In the case of 20, $m = 4$, $a_4 = a_2 = 1$, and $a_3 = a_1 = a_0 = 0$. The binary representation of 20 is therefore 10100.

For $n = 77$, m is 6 and $a_6 = a_3 = a_2 = a_0 = 1$, while $a_5 = a_4 = a_1 = 0$. The binary representation of 77 is correspondingly 1001101.

There are 2^m binary strings of length m, since each digit can be chosen in one of two ways, and for each of these, the next digit can also be chosen in two ways, and as this selection process ripples through the entire string, the number 2 is multiplied by itself m times.

Each individual string $a_{m-1} a_{m-2} \ldots a_0$ corresponds to a whole number between 0 and $2^m - 1$. The reason why this is so is to be found in the expression ($\star$) above: The integer n is never less than zero (simply choose all the coefficients to be zero), and it can never exceed the value obtained by setting all coefficients to one. In the latter case, however, you get a sum $S_{m-1} = 1 + 2 + 2^2 + \ldots + 2^{m-1}$, and from Appendix A this sum equals 2^m

Appendix B: Binary Notation

− 1. Since each of the 2^m possible strings represents an integer, these must necessarily be the 2^m whole numbers ranging from 0 to 2^{m-1}.

As an illustration of the preceeding paragraph, consider all $2^3 = 8$ strings of length 3. It is evident from (★) that the following correspondence holds between the eight triplet strings and the integers from 0 to 7:

000	0
001	1
010	2
011	3
100	4
101	5
110	6
111	7

Turn now to the the *unit interval,* designated by [0, 1], which is the collection of all numbers x that lie between 0 and 1. The number x is an infinite sum of powers of ½:

$$x = b_1/2^1 + b_2/2^2 + b_3/2^3 + \ldots + b_k/2^k + \ldots . \text{ (★★)}$$

To see how (★★) comes about one engages in a game similar to "twenty questions," in which the unit interval is successively divided into halves, and a positive response to the question "is x in the right half?" result in a power of ½ being added in; a negative reply accrues nothing. More specifically, the first question asks whether x is in the interval (½, 1], the set of all numbers between ½ and 1 not including ½. If so, add ½ to the

Appendix B: Binary Notation

sum, namely choose b_1 to be 1. Otherwise, if x lies within the interval $[0, \frac{1}{2}]$, the set of numbers between 0 and ½ inclusively, put b_1 equal to 0. Now divide each of these subintervals in half again, and repeat the questioning: If b_1 is 1, ask whether the number is within $(\frac{1}{2}, \frac{3}{4}]$ or $(\frac{3}{4}, 1]$. In the first instance let b_2 equal 0, and set b_2 to 1 in the second. The same reasoning applies to each half of $[0, \frac{1}{2}]$ whenever b_1 is 0. Continuing in this manner, it becomes clear that b_k is 0 or 1 depending on whether x is within the left or right half of a subinterval of length $\frac{1}{2}^k$. The subintervals progressively diminish in width as k increases, and the location of x is captured with more and more precision. The unending series of terms in (★★) represents successive halvings carried out ad infinitum.

The binary expansion of x is defined as the infinite binary string $b_1 b_2 b_3 \ldots$.

As an example consider $x = \frac{1}{3}$, which is the sum $\frac{1}{4} + \frac{1}{16} + \frac{1}{64} + \ldots$. This sum can also be written as $\frac{1}{4} + \frac{1}{4}^2 + \frac{1}{4}^3 + \ldots$ and from Appendix A, we see that the latter sum is represented by Σ_0 for $a = \frac{1}{4}$, namely ¼ divided by ¾, or ⅓. The binary expansion of ⅓ is therefore the repeated infinite pattern $01010101\ldots$. Another instance is the number $\frac{11}{16} = \frac{1}{2} + \frac{1}{8} + \frac{1}{16}$, and the binary expansion is then simply $1011000\ldots$ with an unending trail of zeros. For numbers that are not fractions, such as $\pi/4$, the corresponding binary string is less obvious, but the previous argument shows that there is, nevertheless, some sequence of zeros and or ones that represents the number.

A number like $\frac{11}{16}$ can also be given, in addition to the expansion provided above, the binary expansion $1010111\ldots$

Appendix B: Binary Notation

with 1 repeated indefinitely. This is because the sum of this infinite stretch of ones is, according to Appendix A, equal to $\frac{1}{16}$, and this is added to 1010. Whenever two such expansions exist, the one with an infinity of zeros is the expansion of choice.

With this proviso every number is the unit interval has a well-defined binary expansion, and conversely, every such expansion corresponds to some specific number in the interval; *there is therefore a unique identification between every one of the numbers in the unit interval and all possible infinite binary strings.* In technical note 1.2 it is further established that *all infinite binary strings are uniquely identified with all possible outcomes of a Bernoulli $\frac{1}{2}$-process.*

These notions are invoked in several places in the book. A specific application follows below.

For any finite binary string s of length n let Γ_s denote the *interval $[x, x + \frac{1}{2}^n)$ extending from x up to, but not including, $x + \frac{1}{2}^n$, where x is the number in the unit interval corresponding to the terminating string s.* Every infinite string that begins with s defines a number within the interval Γ_s because the remaining digits of the string fix a number $a_{n+1}/2^{n+1} + a_{n+2}/2^{n+2} + \ldots$ that never exceeds the sum $\frac{1}{2}^{n+1} + \frac{1}{2}^{n+2} + \ldots$, and according to Appendix A, the last sum equals $\frac{1}{2}^n$. The length of the interval Γ_s is evidently $\frac{1}{2}^n$.

There are 2^n distinct binary strings s of length n, and they define 2^n evenly spaced numbers in the unit interval ranging from 0 to $1 - \frac{1}{2}^n$. The reason is that the sum (★★) terminates with $\frac{1}{2}^n$, so that if x is the number associated with any of these strings, the prod-

Appendix B: Binary Notation

uct $2^n x$ is now an integer of the form (★). These integers range from 0 to $2^n - 1$, as you know, and so dividing by 2^n, it is not difficult to see that the number x must be in the range from 0 to $1 - \frac{1}{2}^n$.

The corresponding intervals Γ_s are therefore disjoint, meaning that they do not overlap, and their sum is necessarily equal to one, since there are 2^n separate intervals, each of length $\frac{1}{2}^n$.

For example, the $2^3 = 8$ strings of length 3 divide the unit interval into 8 subintervals of length $\frac{1}{8}$. The correspondence between the eight triplet strings and the numbers they define is as follows:

000	0
001	$\frac{1}{8}$
010	$\frac{1}{4}$
011	$\frac{3}{8}$
100	$\frac{1}{2}$
101	$\frac{5}{8}$
110	$\frac{3}{4}$
111	$\frac{7}{8}$

Appendix C: Logarithms

The properties of logarithms used in Chapters 2 and 4 to discuss randomness in the context of information and complexity are reviewed here.

The *logarithm,* base b, of a positive number x is the quantity y that is written as $\log x$ and defined by the property that $b^y = x$. *The only base of interest in this book is $b = 2$,* and in this case, $2^{\log x} = x$.

By convention, a^0 is defined to be 1 for any number a. It follows that $\log 1 = 0$. Also, it is readily apparent that $\log 2 = 1$.

Since 2^y means $\frac{1}{2}^y$, then $-\log x = 1/x$ whenever $y = \log x$.

Taking the logarithm of a power a of x gives $\log x^n = a \log x$. This is because $x = 2^y$, and so $x^a = (2^y)^a = 2^{ay} = 2^{a \log x}$. For example, $\log 3^4 = 4 \log 3$.

Appendix C: Logarithms

Whenever u is less than v, for two positive numbers u, v, the definition of logarithm shows that $\log u$ is less than $\log v$.

It is always true that $\log x$ is less than x for positive x. Here are two examples to help convince you of this fact:

Log $x = 1$ when $x = 2$.

When $x = 13$, then $\log 13$ is less than $\log 16$, which is the same as $\log 2^4$, and as you saw above, this equals 4 log 2, namely 4. Therefore, $\log 13$ is less than 13.

Let u and v be two positive numbers with logarithms $\log u$ and $\log v$. Then the product uv equals to $2^{\log u} 2^{\log v}$, which is the same as $2^{(\log u + \log v)}$. It now follows that $\log uv = \log u + \log v$. A similar argument establishes that the logarithm of a product of k numbers is the sum of their respective logarithms, for any positive integer k.

Here is a simple application of logarithms that appears in Chapter 4:

From Appendix B it is known that an integer n may be written as a finite sum $a_k 2^k + a_{k-1} 2^{k-1} + \ldots + a_1 2^1 + a_0$, for some integer k, where the coefficients a_1 are either 0 or 1, $i = 0, 1, \ldots, k-1$, and $a_k = 1$. This sum is always less than or equal to the sum obtained by setting all coefficients a_i to 1, and by Appendix A, that equals $2^{k+1} - 1$, which, in turn, is evidently less than 2^k. On the other hand, n is never less than the single term 2^{k+1}. Therefore, n is greater than or equal to 2^k and less than 2^{k+1}. Taking logarithms of 2^k and 2^{k+1} shows that $\log n$ is no less than k but less than $k + 1$. It follows that the number of bits needed to describe the integer n in binary form is, roughly, $\log n$.

References

Note: The items marked with ★ are especially recommended to the general reader.

1. Ascioti, A., Beltrami, E., Carroll, O., and Wireck, C. *Is There Chaos in Plankton Dynamics?* J. Plankton Research 15, 603–617, 1993.
2. Attneave, F. *Applications of Information Theory to Psychology,* Holt, Rinehart and Winston, 1959.
3. Ayer, A.J. *Chance,* Scientific American 213, 44–54, 1965.★
4. Bak, P. *How Nature Works,* Springer-Verlag, 1996.★
5. Bak, P., Chen, K., and Wiesenfeld, K. *Self-Organized Criticality,* Physical Review A 38, 364–374, 1988.

References

6. Bar-Hillel, M., and Wagenaar, W. *The Perception of Randomness,* Advances in Applied Mathematics 12, 428–454, 1991.
7. Bartlett, M. *Chance or Chaos?,* J. Royal Statistical Soc. A 153, 321–347, 1990.
8. Bennett, D. *Randomness,* Harvard University Press, 1998.★
9. Bennett, C. *Demons, Engines, and the Second Law,* Scientific American 255, 108–116, 1987.★
10. Bennett, C. *Logical Depth and Physical Complexity,* in *The Universal Turing Machine: A Half-Century Survey,* Rolf Henken, editor, Springer-Verlag, 1995.
11. Bennett, C. *On Random and Hard-to-Describe Numbers,* IBM Report RC-7483, 1979.
12. Bernstein, P. *Against the Gods,* John Wiley, 1996.★
13. Borges, J. *Labyrinths,* New Directions, 1964.
14. Casti, J. *Truly, Madly, Randomly,* New Scientist, 32–35, 1997.★
15. Cawelti, J. *Adventure, Mystery, and Romance,* University of Chicago Press, 1976.
16. Chaitin, G. *Randomness and Mathematical Proof,* Scientific American 232, 47–52, 1975.★
17. Chaitin, G. *Information-Theoretic Computational Complexity,* IEEE Trans. on Information Theory IT-20, 10–15, 1974.
18. Chandrasekhar, S. *Beauty and the Quest for Beauty in Science,* Physics Today, 25–30, 1979.
19. Collins, J., and Chow, C. *It's a Small World,* Nature 393, 409–410, 1998.★

References

20. Cover, T., and Thomas, J. *Elements of Information Theory,* John Wiley, 1991.
21. Daub, E. *Maxwell's Demon,* Stud. Hist. Phil. Sci 1, 213–227, 1970.★
22. David, F. *Games, Gods, and Gambling,* Hafner Publishing Co., 1962.★
23. Davis, M. *What Is a Computation?,* in *Mathematics Today,* edited by Lynn Steen, Springer-Verlag, 1978.★
24. Ehrenberg, W. *Maxwell's Demon,* Scientific American 217, 103–110, 1967.★
25. Ekeland, I. *The Broken Dice,* University of Chicago Press, 1993.★
26. Falk, R., and Konold, C. *Making Sense of Randomness: Implicit Encoding as a Basis for Judgment,* Psychological Review 104, 301–318, 1997.★
27. Feller, W. *An Introduction to Probability Theory and its Applications,* Volume 1, Second edition, John Wiley, 1957.
28. Ford, J. *How Random is a Coin Toss?,* Physics Today 36, 40–47, 1983.★
29. Freedman, D., Pisani, R., Purves, R., and Adhikari, A, *Statistics,* Third Edition, Norton, 1998.
30. Frost, R. *Collected Poems,* Penguin Putnam, 1995.
31. Gardner, M. *The Random Number Omega Bids Fair to Hold the Mysteries of the Universe,* Scientific American 241, 20–34, 1979.★
32. Gell-Mann, M. *The Quark and the Jaguar,* W.H. Freeman, 1994.★

References

33. Gould, S. *The Evolution of Life on the Earth,* Scientific American, 85–91, 1994.★
34. Gould, S. *Eight Little Piggies,* Norton, 1993
35. Hacking, I. *The Taming of Chance,* Cambridge University Press, 1990.★
36. Johnson, G. *Fire in the Mind,* Vintage Books, 1996.★
37. Kac, M. *What is Random?,* American Scientist 71, 405–406, 1983.★
38. Kac, M. *Probability,* Scientific American 211, 92–108, 1964.★
39. Kahn, D. *The Code-Breakers,* Macmillan, 1967.
40. Kahneman, D. and Tversky, A. *Subjective Probability: A Judgment of Representativeness,* Cognitive Psychology 3, 430–454, 1972.
41. Kauffman, S. *At Home in the Universe,* Oxford University Press, 1995.★
42. Knuth, D. *Seminumerical Algorithms,* Volume 2, Addison-Wesley, 1969.
43. Leibovitch, L. *Fractals and Chaos,* Oxford University Press, 1998.★
44. Li, M., and Vitanyi, P. *An Introduction to Kolmogorov Complexity and Its Applications,* Second Edition, Springer, 1997.
45. Lucky, R. *Silicon Dreams: Information, Man and Machine,* St. Martin's Press, 1989.★
46. Machlup, S. *Earthquakes, Thunderstorms, and Other 1/f Noises,* in *Sixth International Symposium on Noise in Physical Systems,* 157–160, National Bureau of Standards, 1977.
47. Mandelbrot, B. *The Fractal Geometry of Nature,* W.H. Freeman, 1977.

References

48. Mandelbrot, B., and Van Ness, J. *Fractional Brownian Motions, Fractional Noises and Applications,* SIAM Review 10, 422–437, 1968.
49. Marsaglia, G. *Random Numbers Fall Mainly in the Planes,* Proc. National Academy of Sciences 61, 25–28, 1968.
50. May, R. *Simple Mathematical Models with Very Complicated Dynamics,* Nature 261, 459–467, 1976.
51. Monod, J. *Chance and Necessity,* Knopf, 1971.
52. Newman, J. *The World of Mathematics,* 4 volumes, Simon and Schuster, 1956.*
53. Ore, O. *Cardano, the Gambling Scholar,* Princeton University Press, 1953.*
54. Patch, H. *The Goddess Fortuna in Mediaeval Literature,* Harvard University Press, 1927.
55. Penrose, R. *Shadows of the Mind,* Oxford University Press, 1994.*
56. Pierce, J. *An Introduction to Information Theory,* Second edition, Dover Publications, 1980.
57. Pincus, S., and Singer, B. *Randomness and Degrees of Regularity,* Proc. National Academy of Sciences 93, 2083–2088, 1996.
58. Pincus, S., and Kalman, R. *Not All (Possibly) "Random" Sequences Are Created Equal,* Proc. National Academy of Sciences 94, 3513–3518, 1997.
59. Prigogine, I., and Stengers, I. *Order Out of Chaos,* Bantam, 1984.*
60. Ruelle, D. *Chance and Chaos,* Princeton University Press, 1991.*

References

61. Schroeder, M. *Fractals, Chaos, and Power Laws,* W.H. Freeman, 1991.
62. Shannon, C., and Weaver, W. *The Mathematical Theory of Communication,* University of Illinois Press, 1949.
63. Stewart, I. *Concepts of Modern Mathematics,* Dover, 1995.
64. Voss, R., and Clarke, R. *1/f Noise in Music: Music from 1/f Noise,* J. Acoust. Soc. Am. 63, 258–263, 1978.
65. West, B. and Shlesinger, M. *The Noise in Natural Phenomena,* American Scientist 78, 40–45, 1990.★
66. West, B. and Goldberger, A. *Physiology in Fractal Dimensions,* American Scientist 75, 354–365, 1987.★
67. Zurek, W. *Algorithmic Randomness and Physical Entropy,* Physical Review A, 4731–4751, 1989.

Index

Algorithm, 66, 107, 170
Algorithmic complexity, 3, 92–95, 112, 119–120
Algorithmic randomness, 94–95
Approximate Entropy, 53–54, 56, 63, 81, 137, 164, 176

Bach, Johann Sebastian, 138
Bak, Per, 125, 126, 139, 142, 151
Bar-Hillel, Maya, 29, 147
Bartlett, M., 66, 148

Bartlett's algorithm, 66–69,
Bennett, Charles, 113, 114, 119, 120, 150
Bennett, Deborah, 146
Bernoulli p-trials, Bernoulli p-process, 15
Bernoulli, Jakob, 5–6, 14, 146
Bernstein, Peter, 27, 147
Berry paradox, 108
Berry, G., 108
Binary notation, 13, 183–188
Binary strings, 14
Boltzman, Ludwig, 82, 83

Index

Borel, Emile, 30–32
Borges, Jorge Luis, 2, 92, 100, 150

Caesar, Julius, 3
Calvino, Italo, 1
Cardano, Girolamo, 4, 146
Carroll, Lewis, 65
Cawelti, John, 127
Central Limit theorem, 22, 24
Chaitin, Gregory, xv, 92, 107–109, 149
Champernowne, David, 33, 62, 96, 174–175
Church, Alonzo, 106
Church–Turing thesis, 106
Clarke, John, 138
Coarse graining, 73, 84
Codes, error-correcting, 51
Codes, for encryption, 51–52
Codes, Shannon, 3, 47–48, 94, 99, 163
Conditional probability, 155
Confidence interval, 19–21
Confidence level, 25–26, 61
Crichton, Michael, 35

De Moivre, Abraham, 5–6, 18

De Moivre's theorem, 18–22, 61
Deterministic chaos, 72
Digrams, 49, 54
DiMaggio, Joe, 30
DNA, 121–122
Dovetailing procedure, 111

Elementary event, 11,
Entropy, 40, 42, 87–89, 94, 101, 126

Fabris, Pietro, 128
Falk, Ruma, 63–64, 147
Feller, William, xiii, 145
Fermat, Pierre de, 4, 169
Fortuna, Roman Goddess, ix, 2–4, 146
Fourier, Joseph, 131
Fractals, xvi, 129–131
Frost, Robert, 119, 151

Galton, Francis, 24, 142
Gauss, Carl Friedrich, 6, 22
Gaussian curve. See normal curve.
Gell-Mann, Murray, 122, 151
Geometric sum, 179
Gödel, Kurt, xv, 107

Index

Gould, Stephen Jay, 123, 151
Guare, John, 128

Hacking, Ian, 147
Halting probability, 109–110
Hamilton, William, 128
Heisenberg, Werner, 116
Herbert, Victor, 144
Hurst, Edwin, 136
Hutchinson, G. Evelyn, 124

Information content, 36–39, 56–57, 99
Information theory, 3, 36
Iteration, 66, 167

Jackson, Shirley, ix
Janus sequence, Janus algorithm, 71, 81, 86, 88–90, 98–99, 166–167
Janus, Roman God, xiv, 71, 101
Johnson, George, 87, 149
Joseph effect, 136

Kac, Mark, 146
Kahn, David, 51, 147
Kahneman, Daniel, 64, 147

Kauffman, Stuart, 124–125, 140, 151
Kolmogorov, Andrei, xv, 8, 46, 92
Konold, Clifford, 63–64, 147

Laplace, Pierre-Simon, 34, 64, 115–116, 150
Law of Averages, 16–18
Law of Large Numbers, 5–6, 15–18, 55, 84, 142
Law, 1/f, 139, 141
Leibovitch, Larry, 148
Linus, in comic strip "Peanuts," 29
Logical depth, 120
Lucky, Robert, 50, 147

Machlup, Stephan, 138
Mandelbrot, Benoit, 129, 151
Marsaglia, George, 81, 169
Maxwell, James Clerk, 82
Maxwell's demon, 82, 148
May, Robert, 170
Measure zero, 159
Monod, Jacques, x, 117, 123, 151
Mutually exclusive events, 12

Index

Noises, 1/f, 134–136, 139
Normal curve, 6–8, 20
Normal law, 22
Normal number, 31–33, 59, 62, 95–96, 173
Null hypothesis, 25

Paradox of the plankton, 124
Pascal, Blaise, 4
Patch, Howard, 2
Penrose, Roger, 149
Persistence and anti-persistence, 136
Pincus, Steve, 148
Pink noise, 135, 138, 141
Poincaré, Henri, 24
Prefix code, 47, 94
Prigogine, Ilya, 126, 140, 151
Probability theory, xi, 5, 8–12, 24, 154–155
Program, 102, 104, 106
Pseudo-random, 80

Quetelet, Adolphe, 8

Random number generator, 78–79, 81

Random process, 13
Runs test, 28

Sample average, 5, 16, 21
Sample space, 10,
Santa Fe Institute, 126
Second Law of Thermodynamics, 84, 87–88, 99, 126
Self-organized behavior, 125, 142
Shannon, Claude, xiii, 36, 38, 46, 47, 49, 50
Solomonoiff, Ray, 149
Spectrum, 133–134, 137, 176
Statistical independence, 12, 30, 154
Statistics, 5, 8–9,
Stewart, Ian, 159, 168
Stoppard, Tom, 118
Strogatz, Steven, 128
Strong Law of Large Numbers, 30–32, 77, 155
Swift, Jonathan, 143, 151
Symbolic dynamics, 68
Szilard, Leo, 83, 87
Szilard's demon, 83, 86, 88, 99, 126, 148

Index

Trigrams, 50, 55
Turing computation, Turing machine, 102–105, 171–172
Turing program, 107–108
Turing, Alan, xv, 102
Tversky, Amos, 64

Uniformly distributed events, 12

Vernam, Gilbert, 51
Voss, Richard, 138, 152

Wagenaar, Willem, 29, 147
Watts, Duncan, 128
White noise, 134, 138
William of Occam, 115

Zurek, W. H., 99, 150